Nilo Nunes

Itatinga Ecotourism Theme Park Project - Itatinga Ecopark

Nilo Nunes

Itatinga Ecotourism Theme Park Project - Itatinga Ecopark

Planning and Management of Tourism Enterprises

ScienciaScripts

Imprint

Any brand names and product names mentioned in this book are subject to trademark, brand or patent protection and are trademarks or registered trademarks of their respective holders. The use of brand names, product names, common names, trade names, product descriptions etc. even without a particular marking in this work is in no way to be construed to mean that such names may be regarded as unrestricted in respect of trademark and brand protection legislation and could thus be used by anyone.

Cover image: www.ingimage.com

This book is a translation from the original published under ISBN 978-3-330-75865-0.

Publisher:
Sciencia Scripts
is a trademark of
Dodo Books Indian Ocean Ltd. and OmniScriptum S.R.L publishing group

120 High Road, East Finchley, London, N2 9ED, United Kingdom
Str. Armeneasca 28/1, office 1, Chisinau MD-2012, Republic of Moldova, Europe
Printed at: see last page
ISBN: 978-620-8-30337-2

CONTENTS

ACKNOWLEDGMENTS 2

INTRODUCTION 3

CHAPTER 1 - GENERAL INFORMATION 8

CHAPTER 2 - MUNICIPALITY OF BERTIOGA 11

CHAPTER 3 - ENVIRONMENTAL AND TOURIST CHARACTERIZATION 24

CHAPTER 4 - CHARACTERIZATION OF THE DEVELOPMENT 48

CHAPTER 5 - PROPOSALS 84

CHAPTER 6 - FEASIBILITY AND SUSTAINABILITY STUDIES 91

BIBLIOGRAPHICAL, ICONOGRAPHICAL and PHOTOGRAPHICAL REFERENCES 105

ACKNOWLEDGMENTS

I would like to thank the teachers and staff of the Postgraduate Lato Sensu MBA in Tourism: Planning, Management and Marketing course at the Catholic University of Brasilia, especially **Prof.ª Me. Gladis Lucia Maddalozzo Granemann**, the course coordinator at the time, and **Prof.ª Me Adriana Papaleo**, the supervisor of this work.

To our fellow architects: **Elizabeth de Fâtima Correia, Rafael Magalhâes Nunes, Ana Paula Casa Grande de Oliveira and Nelson Portéro Jr** for their willingness to exchange ideas on the subject and accompany us on several visits to the Itatinga Power Station.

To **Congresswoman Prof. Mariângela Duarte**, for her commitment to the preservation of historical, cultural and environmental heritage and for her work in opening the process for the listing of the Itatinga Power Plant with CONDEPHAAT - the Council for the Defense of the Historical, Archaeological, Artistic and Tourist Heritage of the State of São Paulo.

To the employees of the Itatinga Power Station and all the residents of Vila de Itatinga, who always welcomed us with kindness and attention.

To my wife Beth and my children Daniel, Rafael, Lais Samira and Frederico for their support and encouragement.

Finally, I would like to thank all my colleagues at the **Bertioga Institute of Research and Environmental Sciences - IPECAB**, for their constant support in carrying out my research.

(a) NILO NUNES

INTRODUCTION

This work presents the Project for the Implementation of the Itatinga Ecotourism Theme Park, also known as the **"Itatinga EcoPark"**.

In order to support the future **Preliminary Environmental Report - RAP[1]** , the Project Implementation Roadmap was adapted to the Preliminary Basic Methodology for preparing the RAP.

The project is proposed to use the area of the Itatinga Hydroelectric Power Plant - UHE, with (16,327,070m)[22] , in the municipality of Bertioga - Baixada Santista, on the coast of the State of São Paulo, Brazil.

The municipality of Bertioga is located on the coast of the State of São Paulo, with the **zero point** of the Municipal Cadastral Reference Physical Network, between the geographic coordinates: latitude: 23°51'17"- S and longitude: 46°08'03"- W, forming part of the Metropolitan Region of Baixada Santista (see data in Chapter 1 - General Information).

The municipality of Bertioga is 106 km from the city of São Paulo, capital of the State of São Paulo, 201 km from Campinas, 67 km from Santos (via Cubatâo); 60 km from Mogi das Cruzes, 37 km from Guarujà via the highways or 5 minutes by ferry boat + 28 km from Santo Amaro Island and 100 km from São Sebastiao; Access **from the capital, São Paulo,** via the Anchieta Highway (SP-150) and **the** Imigrantes Highway (SP-160); **from the north coast**, via the Rio-Santos Highway (BR-101) and **the** Dr. **Manoel Hyppólito** Highway (SP-160). Manoel Hyppólito Rego Highway (SP-55); from the **south coast**, via the Padre Manoel da Nóbrega Highway (SP-55) and, **from the plateau**, in Mogi das Cruzes, via the Dom Paulo Rolim Loureiro Highway - Mogi/Bertioga Road (SP-98).

Access to the town and the development site can also be made by sea and estuarine navigation, via the Bertioga Canal; and by river navigation on the Itapanhaù River.

The following **figures** show the location of the area in relation to Brazil, the State of São Paulo, the Metropolitan Region of Baixada Santista and the municipality of Bertioga.

---

1 **RAP**: Abbreviation for "Relatório Ambiental Preliminar" (Preliminary Environmental Report), a document provided as one of the forms of presentation for Environmental Studies. Resolution No. 237, of 19.12.1997, of the National Environment Council - CONAMA.
2 **VIDE TABLE 1 - AREAS:** Areas calculated on the basis of the General Plan: "Terrenos do Porto de Santos na Regiâo de Bertioga", Reference No. I-VII - 11188, dated 17.03.1995. Scale **1:20.000** - CODESP.

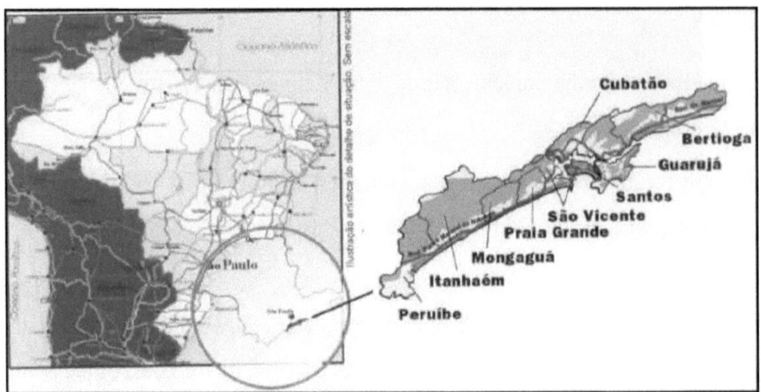

Location of the municipality of Bertioga in relation to Brazil, the State of São Paulo and the Metropolitan Region of Baixada Santista (IPECAB Collection)

Location of the Itatinga plant in the municipality of Bertioga in relation to the Port of Santos (Source: Geribello, 2016).

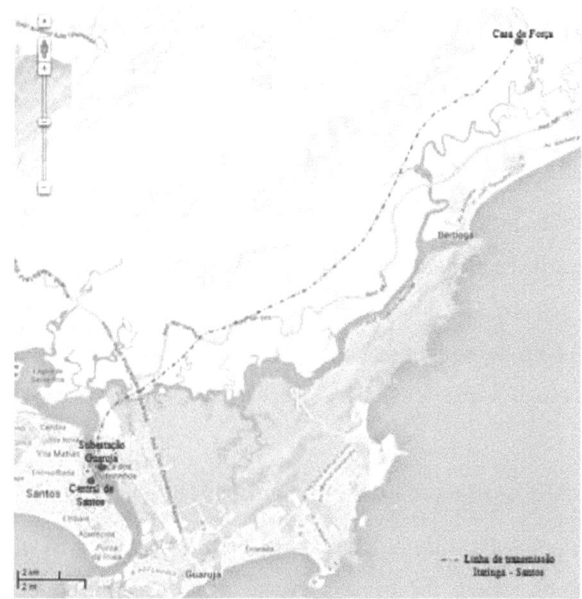

São Vicente Island (Santos and Sâo Vicente), Santo Amaro Island (Guarujà), Bertioga Channel, Itapanhaù River, and location of the Itatinga Power Station Power House (Source: Geribello, 2016)

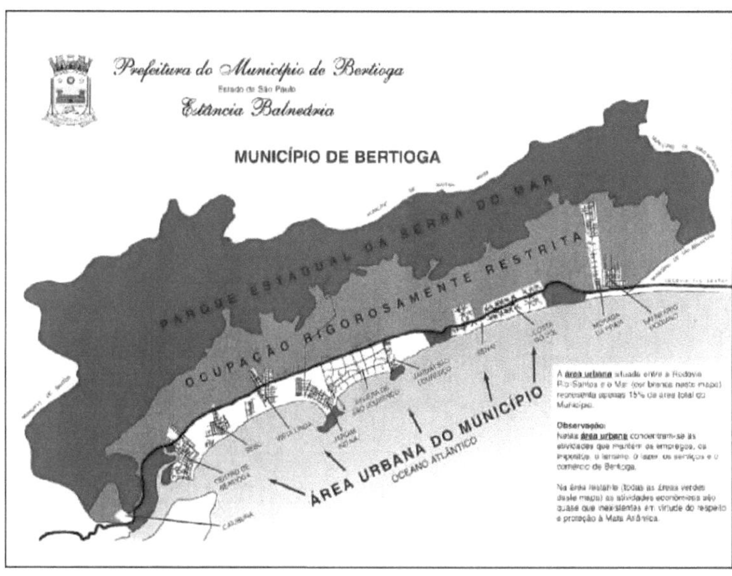

Bertioga municipality area

The definition of the **Park Area**, located in the municipality of Bertioga, includes a **Gate** and

Reception, in an area adjacent to the main access road to the Port of Embarkation (A) for crossing the Itapanhaù River, using an appropriate boat.

Area of the river crossing (coming and going + mooring and maneuvering).

Area of the "Itapanhaù" **Station (1),** where the "Bondinho" is boarded.

Area of the "cable car" route to Vila de Itatinga, which includes a 10 (ten) meter wide strip, **areas of ruins** (archaeological sites), **intermediate areas** along the "cable car" route (mainly stream crossings and observation points on the slopes of Serra do Mar), the **Itatinga Village area**, including the Station (2), with **all the equipment, areas of trails, routes, viewpoints, catchment, dam, adductor channel, water chamber, winches, preservation areas, protection, etc.**

The **total area of the park** [see TABLE 1] corresponds to the sum of all these areas.

It is worth noting that in the design of the park, in order to consolidate a general vision, organize the formulation of proposals, identify the thematic content, define uses, dimension support capacity, identify risks, propose equipment, preserve the environment and other actions, a *preliminary zoning* was established, delimiting spaces with specific objectives and characteristics.

TABLE 1 = AREAS

LOCAL	AREA (m)^2
Portal and Reception	160
Parking Yard	3.000
Port of shipment (A)*	600
Crossing Dominion Strip (coming and going + docking and maneuvering)	3.000
Port of Shipment (B)**	150
Station (1): Cable car (with boarding platform)	60
Cable car route	70.000
Archaeological Ruins	100
Vila de Itatinga (including Station 2)	250.000
Others: Itatinga River Dam, Adductor Canal, Water Chamber, Power Plant Adductor Pipes, Trails, Winches, Preservation and Protection Areas, etc.	16.000.070

TOTAL PARK	16.327.070

* Port (A) = Access Port to the Park, located on the left bank of the Itapanhaù River.

** Port (B) = Disembarkation Port attached to Station (1) of the Cable Car, located on the right bank of the Itapanhaù River.

Note: The right or left bank is considered to be from the source to the mouth.

CHAPTER 1 - GENERAL INFORMATION

1.1　Owner of the area:

Companhia Docas do Estado de Sao Paulo:

CODESP - Administrator and Port Authority of the Port of Santos.

Headquarters address (complete): Avenida Rodrigues Alves, s/n° - Bairro do Macuco.

Santos (SP) - Brazil - CEP = 11.015-201. TEL (13) 3233-6565.

Project Address: Itatinga Hydroelectric Power Plant - Municipality of Bertioga (SP).

1.2　Possible partnerships:

* Bertioga City Hall.

* IPECAB - Bertioga Environmental Sciences and Research Institute.

* CODESP - Cia Docas do Estado de Sao Paulo.

* AGEM - Metropolitan Agency of Baixada Santista.

* Secretariat of Tourism of the State of Sao Paulo.

* Ministry of Tourism.

1.3　Author of the project

- Project author: **Nilo Nunes.**

Architect and Urban Planner - MBA in Tourism Planning, Management and Marketing (UCB/Brasilia)

Specialist in Regional Planning (FAU/USP) Master's Degree in Urban, Rural and Regional Planning (UNITAU/CNDU)

CAU/BR A2942-4.

1.4　General Data on the Baixada Metropolitan Region

Santista - RMBS

* The criteria for the formation of a metropolitan region that have been established by state legislation basically consider the following: high demographic density, significant conurbation, socio-economic integration, urban and regional functions with a high degree of density and specialization.

- The RMBS is made up of (nine) municipalities: [1]Bertioga,[2] Cubatao,[3] Guarujà,[4] Itanhaém,[5] Mongaguà,[6] Peruibe,[7] Praia Grande,[8] Santos and[9] Sao Vicente.

- Total area: 2,420.50 km2 [EMPLASA/2017]

- Total population $\cong$ 1.8 million inhabitants[IBGE 2017/EMPLASA].

- Boundaries: West and North: Juquitiba, Sao Paulo, Sao Bernardo do Campo, Santo André, Salesópolis, Mogi das Cruzes, Biritiba Mirim, Itariri and Pedro de Toledo.

East and South: Iguape and the Atlantic Ocean.

Complementary Law No. 760 of August 1, 1994, established guidelines for the Regional Organization of the State of São Paulo. Article 1 sets out the objectives:

I - Planning.

II- Cooperation, decentralization and better use of public resources.

III- Rational use of the territory, natural and cultural resources, environmental protection, control over the implementation of public and private undertakings.

IV- Integration and execution of public functions of common interest: planning and land use, transportation and road system, housing, basic sanitation, environment, economic development and social assistance, including health and education.

V- - Reducing social inequalities.

The region is made up of 161 km of beaches and a wide range of tourist attractions and potential, involving natural, historical, cultural, architectural, archaeological, environmental and sporting aspects, attracting tourists from all over the state of São Paulo and Brazil. In addition to the Cubatao Steel and Petrochemical Industrial Park and the Santos Port Complex - the largest and most important in South America, the region has a strong presence in various activities, such as foreign trade - import and export, natural gas and oil from the Santos Basin, Transportation and Logistics, etc.

The Baixada Santista Metropolitan Region was created and regulated by State Complementary Law No. 815, on July 30, 1996, by then Governor Màrio Covas, and was the first in Brazil under the 1988 Federal Constitution. The region is made up of a Deliberative Council - CONDESB, a Metropolitan Fund and the Baixada Santista Metropolitan Agency - AGEM, whose function is to plan and execute public functions of common interest to the

Metropolitan Region.

Metropolitan Region of Baixada Santista (Source: IGC/SP)

CHAPTER 2 - MUNICIPALITY OF BERTIOGA

2.1 Historical and Cultural Aspects

2.1.1 Etymological issues:

There are various interpretations of the etymological origin of the word "Bertioga". The word "Bertioga" was modified over the years from the Tupi word "Buriquioca".

Research by the municipality's Department of Culture, carried out in 1993 by Adamastor Sttoffel, states that there are several currents, such as the studies of Leonardo Arroyo, who accepts the relationship of "Buriquioca" with the mosquito barigui, birigui or biriqui, from the *Tupi mberu'wi*, "small fly, mosquito", an insect abundant in the region.

Adamastor Sttoffel explains that according to the evolution of words, the word would undergo the following transformation: biriqui-oka (oka, from the Tupi: hut, house of the Indians), then biriqui-oka would be "house of flies" or "place of mosquitoes" and would become beriqui-oca, bertioca and, finally, Bertioga.

Hans Staden, a German artilleryman who lived in the fort in 1553 (today's Fort São João), recorded the term "Brikioka", according to what he heard from the local Indians, and it came from a small hill, a local landmark (today's Morro das Senhorinhas), which was so called by the Indians, as can be seen in a deed from 1560 in favor of Domingos Garocho. This document refers to the hill as "Buriquioca". From there, we get "Buriqui", a variation of "Muriqui"*, from the Tupi muri 'ki, a* kind of monkey that lived in packs in the woods and gathered on the hill at night. In this second trend, the word "Buriquioca" would mean "monkey dwelling".

A third version is based on the work of the Santos writer Francisco dos Santos, who, in his work on the same subject, states that "Buriquioca" doesn't just mean "home of monkeys"; it means "home of big monkeys".

According to the writer Francisco dos Santos, for Teodoro Sampaio, the word Buriqui is a corruption of "Myra-Qui" - people who waddle, who come and go. However, the writer T. Sampaio stated that Bertioga originated from "Paraty-Oca" - the refuge or home of mullets, not the home of monkeys. This version is strengthened by the natural conditions of the old Bertioga River, now the Bertioga Channel, considered by the ancient caiçaras to be the spawning ground of the paratis, from the *Tupi pira'ti* - 'white fish', abundant from the coasts of Africa to the coasts of Brazil, white in color and distinguished from mullets by the absence of stripes on the body. The abundance of this fish in the Canal (formerly the Bertioga River) would have led the Indians to call the region pira'ti-oka, "place of the paratis or mullets".

We have taken advantage of this and inserted a fourth possibility into these etymological speculations: the Tupi word *mburi' ti*, which means *palm tree*. In this sense, Bertioga could have evolved from "*mburi'ti-oka*" - "place of palm trees", which are currently identified by various names: coconut palms, coqueiro-buriti, buritizeiro, muriti, muritim and edible *palm trees or "juçara palm trees"*, which exist in large quantities in the region. Even among the diversity of palm trees we have the Indaià, from the *Tupi ini'yà*, "fruit of threads", a common name for several palm trees that live in compact communities and originated the name of one of the neighborhoods and beaches in the municipality, Praia do Indaià, whose palm trees inspired the verses of the doctor and poet "Martins Fontes".

2.1.2 Bertioga, from the indigenous to civilization: five centuries of history

Bertioga is linked to the origin of Brazil as an important geographical point for the first settlements. It is a landmark with centuries-old roots. Its origins can be traced back to the middle of the 16th century due to its privileged position at the entrance to the bar, making it an important strategic point in the defense of Vila de São Vicente.

In 1531, when Martim Afonso de Souza, appointed governor-general of the coast of Brazil, landed in the backwaters of the mouth of the "Rio Bertioga", the site of "Buriquioca", he ordered the construction of the *first palisade* for military defense, which was later transformed into a fortification.

Historical records show that the first contact between Martin Afonso's Armada and Joao Ramalho, who lived among the Indians on the edge of the countryside, in the region above the Serra do Mar, took place at the "Buriquioca" site.

After this conversation, Martim Afonso left some men on land and then headed south (Iguape and Cananéia), to return later and penetrate the bar of Santos and the famous river of São Vicente (estuary of Santos) on January 21, leave and go down to earth the next day, heading to the other side of the island, where there was *already* a small settlement and found the village of São Vicente - the first official village in Brazil, on January 22, 1532.

At the time of the first Portuguese arrival, it is likely that Joao Ramalho left his companion Diogo de Braga in "Buriquioca", a character of unknown origin who seemed to live with him among his Indians and servants. Diogo de Braga was married to an Indian woman and they had lived in the region for many years before the arrival of Martim Afonso and spoke Tupi correctly.

Diogo de Braga and his five sons, plus a few companions left by the governor and grantee, are credited with forming the first village at the site of Buriquioca and building a sturdier stockade (the origins of today's Fort Sao Joao - a National Heritage Site - IPHAN), the oldest fortress in the country, which determined the ownership of the place.

From January 1532, the lands of Vicentina (Peruibe to Rio de Janeiro) were populated, with the Sitio de "Buriquioca" as a true milestone in the colonization of Brazil.

There was partial possession of the sea: Cubatao, Piaçaguera, Cabraiaquara, Titiguapara, Peruti, Itapema, Ilha Pequena, Cabuçu, Itapanhaù, Bertioga and the entire island of Guaibé, exercised by the great figures of colonial occupation: José Adorno, Francisco Adorno, Pascoal Fernandes, Jorge Ferreira, Domingos Pires, Pires Cubas, Pero de Góes, Ruy and Francisco Pinto and many others from the first phase in Vicentina, who carried out its first farming and its first industry. Bertioga was a tangent starting point, because beyond it there was an immense emptiness; and, further north, the backlands, the virgin land, the dominion of the Tamoios and the Tupinambàs, who planted their threatening legend there.

It seems to us, and we don't think it's a false interpretation, that almost everything in the period of colonization, especially in the vicinity of the Vicentina colony - the existence of Vila de Sao Vicente, the birth of Santos, the founding of Sao Paulo and the expansion of Brazil - stemmed from that vanguard of the sea and rearguard of the plateau, its facts and its people, linked to the walls of Sitio de Buriquioca, later Bertioga.

2.2 Aspects of Urban Evolution: from the Sambaquis to the present...

2.2.1 A retrospective of the town's occupation:

At least 2000 years B.C. primitive tribes lived in the region and left behind SAMBAQUIS - piles of shells of various shapes and sizes - as a testimony to their existence. When the Portuguese arrived in Brazil, they encountered various indigenous groups. The Tupis lived in this region.

In 1531, Martim Afonso de Souza set sail from his ships in front of the welcoming waters of Buriquioca and, on January 22, 1532, founded the town of Sao Vicente.

The donation of the Bertioga sesmaria was made to Jorge Pires on January 12, 1545.

According to the notes of Hans Staden - a German artilleryman who was shipwrecked off the coast of Brazil and lived in the fort in 1553 - the fort was built in stone and lime in 1547 by five Mameluk brothers, sons of the Portuguese Diogo de Braga, a companion of Joao Ramalho who lived here before the arrival of Martim Afonso's expedition.

In the 17th and 18th centuries, whaling intensified in Bertioga, resulting in the construction of a *Whale Armory*, which extracted oil for public and private lighting in the towns of Santos, Sao Vicente and Sao Paulo. The *Armação*'s central administration was installed in the old Fort Sao Felipe, now in ruins and built almost on the tip of Guaibê Island (now the municipality of Guarujà, on the opposite side to the town of Bertioga), belonging to the old Buriquioca site, known as "Rabo do Dragao".

To this day, it is possible to identify the remaining port of approximately 200 meters and the remains of the large stone tanks built to store the whales' oil.

Whales were abundant along this stretch of coast, along with the fishing industry and the production of oil, which was also sent to Rio de Janeiro. Although the decline of these activities in Brazil occurred in the 18th century, oil production from *the Bertioga Whale Trap* continued until around 1820.

When the whale oil industry was shut down, Bertioga was reduced to a simple fishermen's village, a resting point for small coastal shipping between Santos and the ports on the north coast of the state of São Paulo. This situation lasted for most of the 19th century.

At the beginning of the 20th century, exactly in 1910, Bertioga was inaugurated on the slopes of the Serra do Mar, the Itatinga Hydroelectric Power Station, which began to supply electricity to the Port of Santos.

Until the end of 1930, access to Bertioga was by boat via the Bertioga Canal from the Santos Estuary. At this time, industrialist José Ermirio de Moraes, who had bought Sitio Indaia at the end of Enseada beach in Bertioga, opened a road from Perequê beach in Guaruja to the Bertioga Canal. In the 1940s, this road was maintained by the Guaruja Town Hall and, in the 1950s, by the State Department of Highways (D.E.R.).

In 1954, the *ferryboat* service was set up, connecting this road to the town of Bertioga, consolidating road access and facilitating urban occupation.

The improvements were driven by the establishment in 1948 of the SESC Summer Camp in Bertioga, the first of the Social Service of Commerce (SESC) system, built as a driving force behind the development of tourism and a milestone in Bertioga's recent history.

The urbanization of Bertioga began at the end of the 1940s with a project for 1,700 plots. But it was in the 1960s that Bertioga began to have basic services such as water and electricity. Although the Itatinga Hydroelectric Power Station had been in operation since 1910, the electricity generated was entirely directed to the Port of Santos, which at the time was managed by the Cia Docas de Santos, so electricity only arrived 50 years later, and it was only in 1965 that Bertioga began to enjoy electricity supplied by BELSA - Bandeirante de Eletricidade S.A, later CESP and currently ELEKTRO.

When the 1970s began, Bertioga already had more than 10,000 plots, representing almost a third of the total number of plots estimated at the beginning of this century (around 70 subdivisions and 33,000 plots).

The 1970s and 1980s saw a surge in urban occupation, mainly due to the construction of the Santos - Rio and Mogi das Cruzes - Bertioga highways.

2.3 Physical and territorial aspects

2.3.1 Area, Extension and Population:

With a total surface area of 491.23 km^2 , the second largest municipality in terms of area among the others in the Baixada Santista Metropolitan Region, Bertioga has 20.4% of the regional area. According to the IBGE/Cidades estimate[2016] , Bertioga has a population of 57,942 inhabitants, of whom 28,976 men and 28,966 women. The demographic density (inhabitants/km^2) = 117.97.

- Number of households (Prefeitura/2010) = 44,834.

- Municipal Human Development Index - IDHM/2010= 0.730 (Source: Atlas Brasil 2013 - United Nations Development Program).

- Voters in 2016 (TSE) = 41,478.

- Length of beaches = 33.1 km (33,100 meters).

- Aquatic limit[3] = 46.15km (46,150 meters).

- Beaches and approximate lengths:

• Cove = 12,000 meters (12 km).

• Sao Lourenço = 5,500 meters (5.5 km).

• Itaguaré = 2,900 meters (2.9 km).

• Guaratuba = 8,000 meters (8.0 km).

• Boracéia (to the municipal border) = 4,700 meters (4.7km).

2.3.2 Water Resources:

- Water Resources = Bertioga belongs to the Baixada Santista Hydrographic Basin and has three sub-basins:

- **Itapanhaù River** - Drainage area = 261.5 km .[2]

The springs form in Mogi das Cruzes and Biritiba Mirim, high up in the Serra do Mar. The river flows into the Bertioga channel. Main tributaries: Rio Itatinga and Rio Jaguareguava.

- **Guaratuba River** - Drainage area = 128.7 km^2 . It rises at the headwaters of the Serra and flows into the sea at Guaratuba beach, near Itaguà hill.

- **Itaguaré River** - Drainage area = 85.3 km^2 . It rises on the slopes of the Serra and flows into the sea at Itaguaré beach.

The rivers that originate in the Serra do Mar have different hydrological regimes according to their stretches. At their sources, in a steep topographic region, they have a torrential regime, with a high capacity for transporting solids during the summer rains. When they reach the low slope coastal plain, their water regime changes from torrential to fluvial.

This change imposed by the topography reduces the capacity for transporting solids, resulting in sedimentation on the stretch of plateau, forming meanders and holes caused by constant siltation and erosion. There is also an intermediate hydraulic region, located in the transition between the foothills and the coastal plain, which has deposited: sands, pebbles, rocky substances, gravel and pebbles.

---

3 **AQUATIC LIMIT:** This is defined as the waterline from the border between the municipality of Bertioga and Santos, Caruara/Caiubura - on the Bertioga Canal, to Boracéia Beach, on the border with the municipality of Sao Sebastiao.

The Itapanhaù and Itatinga rivers have an average width of 10 meters in their upper course, while the Itaguaré and Guaratuba rivers in a similar stretch have an average width of 5 meters.

The Itatinga River, which is directly connected to the park, is the main right-bank tributary of the Itapanhaù River, with a basin of approximately 104 km^2, belonging to the Itapanhaù River Hydrographic Basin (261.5 km^2).

2.3.3 Basic sanitation:

The municipality of Bertioga is not yet fully covered by a sewage collection system. With the exception of the Riviera de Sao Lourenço development, located on Sao Lourenço beach, and the SESC vacation camp[4] - Bertioga, which have implemented their own collection and treatment systems, and some parts of the municipality's urban area, there is still a lot to be done. SABESP, through the *Onda Limpa (Clean Wave) Program run* by the São Paulo State Government, has been working hard to minimize sanitation problems. According to information from the company, between 2007 and 2012 the Program implemented 44 km of collection networks, 2.2 km of "trunk collectors" and 4.5 km of pressurized lines, a sewage treatment plant and 4,149 household connections.

However, the situation is worrying, because due to various problems (lack of efficient supervision, undersizing of individual household systems, poorly built septic tanks, inadequate drains, lack of landfill on urban plots, "sill level"[5] lower than recommended, lack of maintenance and cleaning of septic tanks, lack of environmental education, etc.), part of the sewage is improperly discharged into the urban drainage system.), part of the domestic sewage is improperly discharged into the urban drainage system and reaches the sea, mainly through the mouths of small streams.

This situation, in addition to compromising the balneability of the beaches in some spots and stretches, creates adverse conditions and obstacles to the tourist and economic development of the city.

This issue requires prioritizing investments in sanitation, especially in the coastal region of the state, which has natural tourist attractions and almost all of whose municipalities are classified as *seaside resorts.*

---

4 **SOCIAL SERVICE OF COMMERCE** - SESC Bertioga
5 **SILL HEIGHT**: Term used to designate the height (altimetric level) of the sill of the main door of the house, in relation to the level of the street bed in front of the plot.

Disordered real estate growth increases the demands for environmental sanitation and requires that the urban planning process be constant and permanent.

As for the supply and treatment of water, the first system to collect and add water was that of the SESC Holiday Camp, on the Guaxinduva stream, with collection on the slopes of the Serra do Mar, at 64 meters above sea level.

Between 1962 and 1977, the Monos River Catchment and Dredging System was built to supply the Petrobras oil pipeline pumping station, at an elevation of 15 meters, with an approximate flow of 10 l/s.

Part of the urban area in Praia da Enseada is supplied by a public network operated by SABESP. This area has three main catchments, in the Furnas and Pelaes streams and the Itapanhaù river near the Mogi-Bertioga highway.

The water supply system has been expanded with new reservoirs and a distribution network. The systems in the Centro, Vista Linda, Indaià and Sao Lourenço neighborhoods are already operating. Boracéia has new projects and works to expand services.

The main private water supply is from the Riviera de Sao Lourenço, which draws its water from the Itapanhaù River. All the homes in the Riviera de Sao Lourenço development are served by this system.

The water destined for the Riviera de Sao Lourenço is collected at the headwaters of the Itapanhaù River. From this point, a 4.4 km pipeline takes the water to the Treatment Plant - ETA, which is exclusive to the development. The Riviera WTP currently has the capacity to treat 1.2 million liters/hour. Treatment is monitored by the WTP's own laboratory. After treatment, the water is distributed to all the properties in Riviera through a network of approximately 68.2 kilometers. It should be noted that the Riviera Supply System has been planned and sized to meet the maximum demographic capacity of the project when it is fully implemented.

This is an example that municipal urban planning should follow, i.e. the release of new real estate developments in the municipality and the licensing of new building works must be compatible with the capacity of the treated water supply systems.

On Guaratuba beach there is a system supplied by the Perequê-Mirim River, which supplies the Costa do Sol, Portal de Guaratuba and Barra de Indiripira subdivisions.

In Praia de Boracéia there is the Morada da Praia subdivision, with a catchment in the Pedra Branca stream, at an elevation of 15 m, and another SABESP catchment on the slopes of the Serra do Mar, behind the Silveira River Indigenous Reserve.

2.3.4 Electricity:

Currently in Bertioga, practically 100% of the population is served by electricity, as all the streets already occupied are served by a distribution network. The current concessionaire is ELEKTRO.

2.3.5 Telephony:

The municipality is served by fixed and cellular telephony through the VIVO company, in practically the entire urban area.

2.3.6 Transportation:

The municipality has an urban public transport system with bus lines through Viaçao Bertioga and several regional and inter-city lines: Bertioga-Sao Paulo, Bertioga-Mogi das Cruzes, Bertioga- Guaruja, Bertioga-Santos, Bertioga-Cubatao and Bertioga-Sao Sebastiao, through the companies Breda - Transportes e Serviços, Translitoral, Ultra S/A, Litorânea and EMTU-Empresa Metropolitana de Transportes Urbanos.

The Bertioga-Guaruja crossing and vice versa, via *ferry*, is operating regularly with the "*Ferry-Boat*" operated by DERSA.

2.3.7 Road system:

According to the Sustainable Development Master Plan (PDDS), Bertioga's road system comprises the following hierarchy:

I - Regional roads

II   - Main Road System

III   - Secondary Road System

The Regional Roads are the state and federal highways that cross the municipality, the BR-101 (Rio-Santos), SP-55 (Rodovia Dr. Manoel Hyppólito Rego) and SP-98 (Rodovia Dom Paulo Rolim Loureiro), with areas of dominion and legal jurisdiction set by the State Government of São Paulo and the Union.

The roads of the Main Road System include the Anchieta Avenue (Main Connection Axis), which has already been partly built and runs longitudinally through the municipality.

Avenida Anchieta runs parallel to the highway complex formed by BR-101 and SP-55, acting as the main urban road for collecting and distributing traffic and, in a parallel direction to the seafront, it connects the Ferry-Boat Terminal to the Indaia neighborhood, at the opposite end of Enseada Beach.

Along with the Avenida Anchieta, some of the marginal roads of the Regional Roads and the Main Through Roads were designed, those that transversely establish the road interconnection between the road complex and the main connection axis.

In addition to the Main Road System, the municipality has a series of avenues and streets for local use, which are the accesses to the subdivisions; together with the cycle and pedestrian paths, they form the Secondary Road System.

2.3.8 Territorial units:

The urban area of the municipality, due to its physical, topographical, morphological and environmental characteristics, can be divided into Physical-Territorial Units as follows:

Area 1 = from the border with Santos (Caiubura-Caruara), to the Itapanhaù River/Bertioga Canal.

Area 2 = from the Itapanhaù River / Bertioga Canal to Canto do Indaià, comprising the entire length of Enseada Beach.

Area 3 = Sao Lourenço Beach (Riviera de Sâo Lourenço and Jardim de Sâo Lourenço).

Area 4 = Itaguaré Beach.

Area 5 = Guaratuba Beach (including the Costa do Sol and Guaratuba subdivisions).

Area 6 = Boracéia Beach, up to the border with the municipality of São Sebastiâo.

All these areas can be characterized as "urban compartments" and will not be explained and detailed, as they are not the subject of this work.

The Itatinga Hydroelectric Power Plant area, located outside the municipality's urban area, has its main land access connected to territorial unit **Area 2.**

The main access to the **Park's Gate and Reception** is via a **cross road to the SP-55 highway, which** belongs to the Highway Complex (BR-101/SP-55), currently called Rua Manoel Gajo.

2.4 Political and administrative aspects

Bertioga, a small fishing village, was incorporated into the municipality of Santos in 1944, through Decree-Law No. 14.334 of November 30, 1944 (State Decree), annexing its territory to the municipality of Santos, at the time still in the category of "Vila de Bertioga".

On May 19, 1991, after a "pro-emancipation" movement by the local community, a referendum was held. Emancipation was approved by a significant majority of voters. Of the 3,925 people who voted, 3,698 were in favor of emancipation, or 94.2%.

On December 30, 1991, the municipality was created by State Law No. 7.664.

On October 3, 1992, elections were held for the positions of mayor, deputy mayor and councillors, consolidating political and administrative autonomy with the inauguration of the elected mayor and councillors on January 1, 1993.

2.5 Social and economic aspects

2.5.1 Education

According to the final results of the 2015 INEP School Census, the municipality of Bertioga has a student population of **11,191**, not including higher education. This total is distributed as follows: 1,519 in Kindergarten, 1,834 in Pre-School, 4,784 in Elementary School, 2,748 in High School, 310 in EJA (Youth and Adult Education - Elementary School) and 276 in EJA (Youth and Adult Education - High School).

With regard to the number of schools, the following were identified: State: 09, Municipal: 28, Private: 11, totaling **48 schools**.

The total number of students enrolled in Special Education is: 250, distributed as follows: 12 in nursery schools, 26 in pre-school, 140 in the initial years of elementary school, 22 in secondary school and 50 in EJA. These figures include the Indigenous Municipal School, located in the Silveiras River Indigenous Village in Boracéia.

There are 355 students enrolled in face-to-face undergraduate courses (SEADE data - State Data Analysis System Foundation/2014). This summary does not take into account students studying in higher education outside the municipality.

2.5.2 Health

The municipality's health sector is subordinated to the Municipal Health and Welfare

Secretariat. The health system is supported by the Mixed Hospital Unit, the Maternity Hospital, the Municipal Emergency Room and the Bertioga Health Center.

In addition to this structure, the municipality has Polyclinics, Basic Health Units, Outpatient Clinics and Health Posts in the main districts of the city. The main services provided by the health structure are: Nutrition and Dietetic Service, Breastfeeding Service, Dental Service, Clinical Analysis Laboratory, Hospital Pharmacy, Surgical Center, Wards and Ambulance Service for aid and removal.

In the area of public health, the municipality is active in health surveillance, epidemiological surveillance, prevention of tuberculosis, leprosy, infectious and contagious diseases, and immunization, among others; it also has a Zoonosis and Epidemic Control Sector.

The Department of Social Development, Labor and Income coordinates social action in the municipality. It is responsible for food security; it participates in activities that stimulate the fight against poverty, with policies to promote workers; food security and the coordination of the Municipal Hostel. [a]It operates in social work, child and adolescent care, psychological care, vocational courses for adults, a shelter for children and the elderly, groups for the third age, drug prevention, etc.

2.5.3 Security:

Security in the municipality is carried out by the State Military Police (Fire Brigade, Highway Police and Environmental Police) and the Civil Police, through the Bertioga Police Station and the Municipal Guard.

2.5.4 Housing:

The issue of housing, a social problem, is directly related to laws, land tenure and use, as well as patterns of behavior and growth in the city. The housing problem is essentially political when it is used for electioneering purposes.

There is often implicit or explicit tolerance on the part of municipal authorities of invasions of public areas, permanent preservation areas and even private areas.

We live with conflicting values, interests and social pressures.

The technical field has solutions for minimizing the problem, but the poor population is almost always used as a "mass of manoeuvre" to enable the interests of the majority of politicians. As long as shacks and substandard housing are exchanged for votes, the problem will not be solved.

Bertioga is seeing a growth in favelas and "invaded areas", often sponsored by people interested in getting votes in the elections.

Society needs to discuss this issue in depth, decide how much it will invest and improve legal and fiscal measures.

Urban housing policies require a break with paradigms, professional planning and management, creativity in building solutions and alternative models for a sustainable and healthy city.

CHAPTER 3 - ENVIRONMENTAL AND TOURIST CHARACTERIZATION

3.1 Climatic aspects

3.1.1 Climate Dynamics

The municipality's climatic characteristics are conditioned by the relief and continentality of the coastal and coastal plain, bordered by coastal massifs, escarpments and hills, and the strong incident solar radiation, mainly due to the latitude.

As for climate dynamics, this is influenced by the presence of regional atmospheric circulation systems, with the **Atlantic Tropical Mass** and the **Atlantic Polar Mass** standing out. The **Atlantic Tropical** Mass is characterized by the warm, humid maritime air mass that dominates South America for most of the year. It is very humid, with summer and winter variations and homogeneous temperatures. Its action defines the climatic classification of the region:

- TROPICAL RAINY CLIMATE -

The **Atlantic Polar Mass** that reaches the coast of Brazil, with its cold and dry origins on the Antarctic Continent, acquires the characteristics of a maritime polar mass when it crosses the Atlantic Ocean. The penetration of this air mass into the continent causes an atmospheric situation of instability, generating persistent torrential rain and heavy cloud cover.

3.1.2 Prevailing winds

The **direction of the prevailing winds** in this coastal region is easterly, but with the entry of cold fronts there is a dominance of southerly winds.

The region also suffers from the so-called FOENN effect. As the cold air moves towards the escarpment of the mountains, it becomes trapped and undergoes adiabatic heating; and as it rises, trying to cross the **mountain wall**, it cools down again, becoming saturated with condensation of its moisture, forming areas of clouds that crown the tops of the Serra do Mar, causing the constant fogs, drizzles and heavy torrential rains.

3.1.3 Precipitation (rainfall)

The region has places with high annual rainfall, reaching an average of **4,597 mm/year** at the rainiest point (Posto Pluviométrico da Represa de Itatinga - DAEE/SP), located above the

Itatinga Hydroelectric Power Plant, in the area of the dam, at an altitude of 720 meters.

In the lowlands, stations located 3 meters above sea level have lower values. Example: "Posto Bertioga" at an altitude of 3 meters indicates an **average annual rainfall of 2,692 mm/year**.

3.1.4 Temperatures

For the Bertioga region, which **is essentially touristy**, in addition to the pleasant temperatures where the **average maximum is 26.8°C** and the **average minimum is 18.9°C**, the **number of rainy days during the month is** very important. On average, this **number of rainy days,** measured over a period of 10 (ten) years, is **14 (fourteen) days**.

3.1.5 Sunshine

In terms of **sunshine,** the **total average was 152.5 hours** in January, compared to **96.8 hours in July**.

3.1.6 Relative humidity

Relative humidity at sea level ranges from 78% (November) to 84% (September), with an **annual average of 81.2%**.

3.1.7 Atmospheric Pressure

Atmospheric pressure ranges from 1,010 millibars in summer to 1,018 millibars in winter.

At sea level, atmospheric pressure is always very high and decreases with altitude.

3.2 Geological aspects

From a geological point of view, the Bertioga region can be divided into two main groups:

- Precambrian crystalline basement.

- Sedimentary cover

3.2.1 The crystalline basement

The **crystalline basement** stretches from the Serra do Mar, reaching the entire coastal plain, where it is covered by a thick layer of sediment, until the coastline crosses the plain. The irregularities in the surface of this bedrock form the various rounded hills and islands close to the coast, present along the entire coastline of the State of Sao Paulo.

It is made up of medium- and high-grade metamorphic rocks (gneisses, migmatites and ophthalmites). These rocks were generated by the remelting of the lower crust during a period of great tectonic and magmatic activity (Brasiliano Cycle).

In addition to these ophthalmites, there are localized granite intrusions.

3.2.2 Sedimentary cover

The coastal sedimentary cover covers the crystalline bedrock throughout the Planicie, from the Serra do Mar to the coastline and inland, with the exception of the points where the bedrock rocks protrude, in the form of **hills** (in the Planicie), **islands** (near the coast) or **rocky coasts**, formed by immense rolled blocks.

This **sedimentary cover** is made up of marine or river sands, clays and organic sediments.

The fluvial sands and clays originate on the mainland and in the mountains, due to the chemical and physical weathering of the primary rocks (chemical attack of the rock by acidic water from surface runoff), and the abrasive action of water and wind. They are then transported by the rivers towards the coast to the plateau, where the transport energy decreases due to the low slope. The rivers spread out and the sediments are deposited.

The soil formed by the action of **weathering** on the continent is a fertile substrate for the development of the vegetation of the Serra do Mar, which provides organic matter that is decomposed and transported in solution and suspension (in a colloidal state) by the rivers. When they come into contact with the chlorides and sulphates present in seawater, these compounds in a colloidal state flocculate and precipitate in the form of very fine organic clay, making up the typical sediment of swamps and mangroves.

As the river approaches the sea, its salinity increases and the organic matter begins to flocculate, forming the mangroves, a nutrient-rich and rather soft soil.

3.2.3 The formation of the Sedimentary Plateau

The Bertioga plateau is basically made up of continental and marine sediments, evolving due to the phenomenon of marine regression (lowering of the sea level) which began approximately 120,000 years ago when the sea was 6 to 10 meters higher than it is today, reaching the foot of the Serra do Mar.

The sediments brought from the continent via the rivers were deposited directly into the sea and mixed with the marine sands. With the return of the sea, a "sandy cord" was deposited by the sea along the entire coastline, at the same time as the covering took place; this material mixed with the sediments brought from the continental plateau.

The various sandy ridges were formed, with those closest to the mountain being completely covered by continental sediments.

Where this sedimentation didn't take place, the marine deposits remained exposed, forming lines parallel to the coastline made up of sand. Thus, former islands were incorporated into the mainland, such as Morro do Indaià, in the corner of Praia da Enseada and others.

Approximately 18,000 years ago, this process of lowering the sea level stopped, possibly even receding a little. As a result, the rivers that ran perpendicular to the newly formed coastal plain were blocked by the sea and forced to run parallel to the coastline, meandering and digging into the sandy deposits.

The layers of varying thickness composed of a dark, almost black, very plastic clay, found in various stretches of the Itapanhaù River, come from the deposit of a large amount of organic matter (leaves, branches, animal remains, seeds, fruit, etc.) which over time has been decomposed and compacted into overlapping sediments.

3.2.4 Geomorphological Analysis

The Coastal Zone, which comprises the area known as the Coastal Plain, has peculiar natural characteristics, involving different **ecosystems**, where the relief takes on great significance, both in terms of landscape and the environment.

The physical environment is highly conditioned by the **compartmentalization** of the relief, both regionally and locally.

The coastal area covered by the coast is represented by beaches, sandbanks, coastlines and the land behind them, resulting from marine, riverine and lagoon activity. In contrast to the flat terrain of the mangroves, beaches and sandbanks, there are the coastal hills and massifs and, in the background, the Serra do Mar and its escarpments, which act as a *real wall at* an altitude of approximately 800 meters.

While the plateaus are of extremely recent origin, made up of sandy and locally clayey and organic sediments and with a strong presence of water, the hills and mountain massifs were generated by tectonics and sculpted over the ages. They have a rigid rock structure, with gneisses and migmatites standing out. The plateau area is home to a variety of soils that help define different plant associations.

On the strip bordering the ocean are the **podzolized** hydromorphic soils, with sandy textural characteristics. In these stretches there is *jundu vegetation*, ranging from grasses to woody plants up to 5 meters high, as well as tree vegetation known as *restinga forest*.

Mangrove vegetation develops towards the back of the cordon, along the Bertioga Channel and the mouths of the rivers, with organic soils.

At the foot of the Serra escarpment, there are alluvial and colluvial soils with a clayey or sandy texture, covered by **tropical forest**.

We can thus classify the following compartments of the coastal plain:

- Plain with beaches, jundu and restingas (sandy marine strands).

- Mangrove plain

- Floodplain

- Fluvial-marine plain, mixed detritics (marine, lagoon and river)

Accompanying the base of the escarpment we will also see some landforms such as: **dejection cones**[6] **, alluvial fans**[7] **, talus deposits** and **colluvial ramps**.

3.3 Aspects of vegetation cover

The municipality of Bertioga, with an area of 491.23 km^2 , currently has around 80% of this area covered by natural vegetation.

From an ecological point of view, this vegetation can be divided into four main categories:

a)   Atlantic Forest

b)   Restinga vegetation

c)   Wetland Plant Formations

d)   Mangrove

3.3.1 Atlantic Forest

The Atlantic Forest originally spread along the Brazilian coast, from the north of Rio Grande

---

6 DEJECTION CONES: same as a cone of debris similar to alluvial fans, but in the form of cones.
7 ALLuvial **LEQUES**: accumulation of alluvial material in the shape of a fan, formed by rocks, gravel, pebbles and sands coming from the torrents of the Serra escarpments.

do Sul to the south of Paraiba.

In the state of Sao Paulo, this formation is located along the Serra do Mar and the Serra de Paranapiacaba, and is partially preserved due to the difficulties of using the soil and the presence of very steep slopes, at altitudes of between 500 and 1000 meters.

Despite the devastation that occurred with the arrival of the Portuguese colonizers in the 16th century, beginning with the predatory extraction of brazilwood for firewood; then with the clearing of the forest for the sugar cane cycle; deforestation for mineral extraction; coffee plantations, raw material for the cellulose industry, etc., the Atlantic Rainforest still offers a wide variety of *ecosystems.*

In some specific parts of the Atlantic Forest remnants, such as the **estuarine - lagoon complex of Iguape - Cananéia** and **Paranaguà**, located in the states of Sao Paulo and Paranà, the levels of biodiversity are considered to be the highest on the planet.

In Bertioga, somewhat due to the difficulties of access, the Atlantic Rainforest remains fairly well preserved and is distributed along the entire Serra do Mar, from the border with Santos to the border with Sao Sebastiao.

This is a species-rich plant formation, with a very high rainfall rate, on poor, shallow soil, in a strongly undulating geomorphology. The interaction of these factors, although adverse to the stabilization of a forest community, has led to the formation and maintenance of this plant complex.

Despite its apparent exuberance, it is a formation with a low concentration of nutrients in the soil, which are basically kept in biotic form, i.e. in the plants themselves.

The **nutrient cycling** process is faster than in similarly poor and deeper soils, such as those in the **Cerrado** biome.

The trees of the Atlantic Rainforest, although large, have their root systems at shallow depths and are vulnerable to natural or man-made disturbances.

Apparently homogeneous, the Atlantic Rainforest is made up of different communities, distinguished by their floristic and/or physiognomic characteristics due to geomorphological, edaphic, hydric and climatic variations that are localized. Examples: hillside forests, highland forests, highland fields, valley bottom (or grotto) forests and riverine forests.

Still in relation to the floristic aspect and the high humidity content of the Atlantic Forest, we highlight the presence of a high number of epiphytic species, especially the Bromeliaceae, Araceae and Orchidaceae families.

The Atlantic Forest is home to species that provide wood, such as Brazilian oak (*Euplassa canthareirae*), red jequitibâ (*Carininana legalis*), white jequitibâ (*Carininana estrellensis*), cinnamon *trees* (*Ocotea* spp), Angelim *(Andira anthelminthica)*, caixeta (*Tabebuia cassinoides*) and others.

Among the species used for medicinal purposes are guaçatonga (*Cassearia sylvestris*), cidreira-do-mato (*Hidyosmus brasiliensis*) and pau-de-alho (*Copaifera landdorffii*).

The juçara or içara palm (*Euterpe edulis*), which is currently a victim of predatory exploitation, is also valuable for food.

In the Bertioga region, the Atlantic Rainforest is distributed along the entire Serra do Mar.

Studies of a phytogeographic approach to the Baixada Santista, developed by *Andrade and Lambert* (1965), indicate an approximate diversity of 170 species in the Atlantic Forest of the Bertioga region, which increases significantly towards Boracéia and the border with São Sebastiâo, reaching 480 species.

3.3.2 Restinga vegetation

It is the most representative form of vegetation in terms of area occupied in the municipality. It is estimated that it accounts for 60% of the vegetation cover. They form sandy plains with plants of arboreal and shrubby size.

According to experts, the concept of restinga is variable and can have at least three different types: the plant formations that cover the Holocene sands from the ocean and can reach the first elevations of the Serra do Mar; the landscape formed by the justamaritimo sand with its overall vegetation; or, the woody and dense vegetation of the inner and flat part after the mangrove, the dunes and the jundu.

Generally speaking, the current predominant concept of "restinga" encompasses all the vegetated formations of the sandy coastal plain: beaches, dunes and forest communities.

Restinga ecosystems are characterized by sandy soils, poor in clays and organic matter.

In general, the amount of organic matter in the soil is higher in places with dense vegetation due to the accumulation *of leaf litter* in these places.

The restinga with its tree and shrub formations reaches an average height of 15 to 20 meters.

Considering the floristic composition of coastal plant communities, it can be said that the restinga is influenced by topography, proximity to the sea, soil conditions and the depth of the water table.

The main families that occur in the restingas are: Leguminosae, Rubiaceae, Bromeliceae, Orchidaceae, Myrtaceae, Gramineae and Compositeae.

It should be noted that in the pre-sierra zone, near the first elevations of the Serra do Mar, it is common for some species from the Atlantic Forest to invade the Restinga Forest and vice versa, constituting in some stretches a true "transition zone".

It should be noted that the restinga is a fragile environment in the face of increasing and constant aggression, such as the clearing of large tracts of vegetation for the installation of allotments, real estate speculation, disorderly urbanization, slumification, invasions of public areas, mineral extraction, etc.

The Bertioga sandbank, however, when compared to other municipalities on the north and south coasts, is the one that still has the greatest territorial continuity, either longitudinally to the beach or transversally to it, which guarantees the presence of the various species and communities that make it up for the time being.

3.3.3 Wetland Plant Formations

These are formations located on alluvial land, subject to flooding.

The shrubby vegetation that occurs in areas subject to periodic flooding is called *vàrzea forest*, which is a relatively low formation with undefined strata.

In Bertioga, this forest is found along some of the main rivers, such as the Itapanhaù, Itatinga, Itaguaré, Guaratuba, Bananal and Ribeirao Vermelho, and is represented by not very extensive areas.

These wetland vegetation formations can contain areas with hydrophilous herbaceous vegetation, basically made up of terrestrial herbs, called swamps and located on constantly waterlogged soils, occurring in small patches along the rivers mentioned. They can also occur in arboreal form, known as **Paludosa Forest**.

3.3.4 Mangrove

The mangrove vegetation in the municipality of Bertioga is distributed along the mouths of the three main rivers: Rio Itapanhaù, Rio Guaratuba and Rio Itaguaré, as well as some woodland on the banks of the Bertioga Canal.

The mangrove swamp is a coastal ecosystem generally as far as the tide can reach.

The main characteristic of this plant formation is the presence of species capable of withstanding adverse conditions, such as excessive salinity, continuous tidal flows, poorly developed substrates and low oxygen content, *determined by the constant saturation of the soil by seawater.*

The plant species of these environments can be found in the root system, which, although shallow, has nutritional balances and plays an important role in sustaining and breathing these plants, through the strut roots and pneumatophores.

In order to survive in saline soils, certain mangrove species need to control the concentration of salt in their tissues. This can be done either by preventing salt from entering through the roots or, in some species, by excreting salt through glands in the leaves. Another way of eliminating salt is to drop leaves with excess salt (Mitsch and Gosselink, 1986).

Another important adaptation ensuring the reproduction of these plants is viviparity, i.e. the zygote, once fertilized, forms the embryo which develops while still attached to the "mother plant", giving rise to a seedling which, once fully formed, detaches (Macnae, 1968). This strategy gives the new plant a better chance of success in anaerobic waters that could prevent it from settling.

In Bertioga, the predominant occurrence of this vegetation is on the banks of the rivers and the Bertioga canal up to where the salinity is felt, and these are called riverine forests.

In terms of flora, Bertioga's mangroves are classified into **three species**:

■ **Red Mangrove** (*Rhizophora mangle*), also: mangue bravo, sapateiro, candapuva, guarariba, catimbó, etc.

Important feature: aerial root system and shallow environment.

■ **Siriùba mangrove** (*Avicennia shaueriana*), also: siriba, seribà, white mangrove, guaperu, black mangrove, yellow mangrove, etc.

Important characteristic: It tolerates high salinities and has two types of roots - one that penetrates the substrate and comes back out, and another that is very branched, only submerged horizontally, spongy and fibrous.

■ **White mangrove** (*Laguncularia recemosa*), also:

manso mangrove, tinteiro mangrove, sapateiro mangrove, sibina, etc.

Important characteristic: Low salinity, but tolerates it better than the red mangrove (*R. mangle*). The white mangrove does not survive in places where the water level fluctuates greatly. The trees can reach up to 20 meters, but on average they are between 4 and 6 meters. Their trunks are fissured.

The mangroves are a system of extreme ecological importance; they constitute a habitat and ecological niche for various species of animals that use the mangroves as a place to protect themselves and breed.

The mangrove also acts as a biological filter and accumulates sediment. It produces organic matter; it protects the larvae and young of many species of fish that spawn there, thanks to the muddy substrate and the intricate system of roots. It is recognized as a *"nursery for marine life"*.

3.4 Faunal Aspects

The municipality retains around 80% of its natural vegetation cover, making it important for the diversity of fauna.

Even allowing for the lack of in-depth studies on the various zoological groups (mammals, birds, reptiles and amphibians) in the municipality, it is possible to state that, when compared to other coastal municipalities, Bertioga is home to a significant portion of the original biological diversity, mainly due to the presence of the Atlantic Forest and the high representation of coastal and estuarine formations.

Considering the four main categories of vegetation already described: Atlantic Forest, Restinga Vegetation, Wetland Vegetation Formation and Mangrove, it is not possible to characterize the faunal assemblage of each of them, with the exception of some groups of animals for which it is possible to list some of the most frequent species in a given environment.

We should also point out that some species move around and can be seen in different vegetation categories.

3.4.1 Mammals:

From a zoogeographic point of view, the municipality of Bertioga is home to mammals widely distributed in the Atlantic Forest domain. Some examples of species:

POPULAR NAME	SCIENTIFIC NAME/ MAIN SPECIES
Nail monkey	*Cebus apella*
Bush dog	*Dusicyon thous*
Hand-stitched	*Procyon cancrivorus*
Furâo	*G. vittata*
Bush cat	*Felis concolor*
Ocelot	*F. pardalis*
Tamandua	*Tamanduà tetradactyla*
armadillo	*Dasypus novencinctus*
Tapir	*Tapirus terrestris*
Queixada	*Tayassu pecari*
Wild mouse	*A. serrensis, Akodon cursor* and *A. nigrita*
Tachyderm rat	*Kannabateomys amblionyx*
Hedgehog	*Coendou prehensilis*
Prea	*Cavia aperea*
Capybara	*Hydrochaeris lydrochaeris*
Cutia	*Dasyrocta azarae*
Paca	*Agouti paca*
Gamba	*D. marsupialis*
Bats	Various species (31)*
Cuica	Various species (14)**

* There are several species cataloged (31) ** There are several species cataloged (14) The number beside refers to the number of species cataloged.

In general, it is not possible to list a mammal fauna that is exclusive to the restinga or mangrove swamps, as the species that inhabit these environments are those commonly found in the Atlantic Rainforest.

You can identify the greater or lesser occurrence of a mammal in the so-called closed (forest) formations, others that are adapted to the more open ones (clearings, capoeiras, less dense restingas, fields, etc.) and those that are adapted to both. There are also cases of animals with terrestrial, riverine and aquatic habitats.

In addition to the destruction of natural habitats, another worrying factor that puts many species at risk is predatory hunting.

In the municipality of Bertioga, along trails in the forest, it is possible to see "blind" spots for catching paca, armadillo, opossum, anteater, peccary and tapir.

Larger mammals are becoming increasingly rare due to frequent human persecution and illegal hunting of wild animals. Examples: tapir (*Tapirus terrestris*), deer (*Ozotocercus bezoarticus*), puma (*Felis pardalis*), ocelot (*F. pardalis*) and jaguar (*Panthera onca*).

3.4.2 Birds:

Some preliminary surveys point to around 200 species of birds likely to occur in Bertioga, with the Tyraniidae family (bem-te-vis, tesoura, irerês) being the most representative, due to its diversity, totaling almost 30 species.

Below is a table listing some of the species likely to occur in the municipality:

POPULAR NAME	SCIENTIFIC NAME/ MAIN SPECIES
Black anu	*Drotophaga ani*
White Anu	*Dromococcyx phasianelus*
Swallow	Various species (3) *
Hummingbird	Various species (9) *
Kingfisher	Various species (5) *
Aracari	*Baillonius bailloni*
Toucan	*Ramphastos dicolorus*
Woodpecker	Various species (5) *
Arapacu	Various species (5) *
Barranqueiro	Various species (3) *
Curutie	*Certhiaxis cinnamonea*
Joao-de-barro	*Funaruis rufus*
Leaf cleaner	*Philydor atricapillus*
Bico-virado	*Xenops rutilans*
Inhambu	*Crypturellas obsoletus*
Macuco	*Tinamus solitarius*
Bigua	*Phalacrocorax olivaceus*

Big Punch	*Ardea cocoi*
White Egret	*Casmerodius albus* and *Egretha thula*
Mallard	*Lirina moschata*
Vulture	Various species (6) *
Sparrowhawk	*B. magnirostris*
Gaviao	*Harpagus diodon*
Tickler	*Milvago chimchima*
Blackbird	*Vanellus chilenses*
Torch	Various species (4) *
Rolinha	*Columbina talpacoti*
Parakeet	Various species (3) *
Maitaca	*Pionus maxiliarii*
Did you know	Various species (5) *

* There are several species catalogued; the number beside it refers to the number of species catalogued.

It should be noted that many of the bird species found in Bertioga are linked to the estuary, representing the last stage of the food webs found in the mangroves. Most of them come to these environments in search of food or to rest during their migrations.

In terms of bird feeding habits, we can say that there are frugivores (feeding on fruit), predators (feeding on other animals), nectarivores (feeding on the sugar-rich nectar produced by flowering plants) and saprophagous or scavenging detritivores (feeding on organic remains).

3.4.3 Reptiles:

There is a low density of individuals, due to the umbrella-like tendency or discreet habits of the species, and often due to the dense vegetation and large amount of *litter* on the ground, which makes it difficult to identify the reptiles.

Below are some reptile species that are likely to occur in Bertioga:

POPULAR NAME	SCIENTIFIC NAME/ MAIN SPECIES
Lizards	Various species (16) *
Snakes	Various species (55) *
Amphisbaenians	Various species (5) *

* There are several cataloged species, the number beside refers to the number of cataloged species.

The lizards include the Tupinambis teguixim and the genus Mabya. The latter, although typical of the forest, seems to have the ability to regulate its body temperature and is often found in thickets and clearings.

As for snakes, the Colubridae family is the most representative.

The few observations do not allow us to generalize about the ecology of these animals. Among the different species likely to occur in Bertioga, there are fossorial forms (which frequent the surface layers of the soil and the lower strata of the vegetation), subarboricolous forms (which frequent all the strata of the vegetation, but descend to the ground) and arboricolous forms (which are rarely seen on the ground).

From an ecological and conservation point of view, reptiles and amphibians should be treated together, as many species of snake are mainly fed by amphibians.

With regard to snakes, it should also be borne in mind that these animals have been heavily persecuted. In general, snakebites occur when a venomous snake is stepped on, touched or when it feels threatened by approaching humans.

With the exception of viperids (jararacas and rattlesnakes) and some elapids (true corals), the vast majority of snakes are harmless to humans.

3.4.4 Amphibians:

The species of anurans (frogs) likely to be found in the Bertioga region are those typical of the Atlantic Forest domain, characteristic of forested environments.

From an ecological point of view, it should be noted that for amphibians the place where they reproduce (spawn and develop tadpoles) is of fundamental importance. In this respect, the vast majority of species need open water, and those that build nests in foam (or other terrestrial type) are rare. This makes the water's edge environments found in Bertioga extremely important, such as marshes, floodplains, paludanous forests and riverside forests.

POPULAR NAME	SCIENTIFIC NAME/ MAIN SPECIES
Toad	Various species (43)
Râ	Various species (32)

* There are several cataloged species, the number beside refers to the number of cataloged species.

3.4.5 Fish:

In general, there are few studies on Bertioga's ichthyofauna.

Luedervaldt (1919) presented the first list of fish species in the region in a study of the mangroves of Santos. More recently, Paiva-Filho (1982 to 1987) studied the ichthyofauna of this region, presenting a list of the most frequent species caught.

Therefore, in view of the aims of this study, only the ichthyological community of coastal environments was considered, not least because these sites, as well as having a significant representation in Bertioga, are important for the development of fishing activities, especially sport fishing. This community can be seen as a strong attraction for tourism.

It should be pointed out that although this analysis is of great scientific interest, it is not possible to characterize the fish communities that inhabit the rivers and streams located at the foot of the Serra do Mar. The high transparency of the water, the stony bottom and the high speed of the current make it difficult to carry out the usual collection procedures, but it is worth highlighting the fact that these environments generally have their own communities with a high degree of endemism.

The proposal for the *Itatinga Ecotourism Theme Park* emphasizes the importance of ichthyofauna and fishing in the treatment of some of the park's attractions and also as a thematic element for environmental education.

Considering the ecological aspect and the difficulty in defining fish that are exclusive to the sea, estuary or river, we present an illustrative list of fish that are likely to occur in Bertioga.

POPULAR NAME	SCIENTIFIC NAME/ MAIN SPECIES
Flounder	*Achiurus Iineatus* and *Anchovy filifera*
Fire tail	Various species (3) *
Sardines	*Anchovy marinii*
Herring	*Anchovia clupeoides* and *Anchoviella brevirostris*
Manjuba	*Anchoviella lepidentole*
Sargo	Various species (3) *
Xareu	*Caranxhippos* sp
Yellow catfish	*Catheorops spixii*
Sea bass	Various species (4) *
Yellow hake	*Cynoscion acoupa*
White hake	*Cynoscion leiarchus*
Carapeba	*Diapterus olisthyostomus*
Sole	*Etropus crossotus*
Carapicu	Various species (3) *

Caratinga	*Eugerres brasilianus*
Cara	*Geophegus brasiliensis*
Black needle	*Hemiramphus brasiliensis*
Seahorse	*Hippocampus reidi*
Moth	*Hoplias malabaricus*
Pie	*Isopsthus parvipinnis*
Freshwater catfish	*Lucipimelodus platanus*
Corvina	*Micropogonias furniari*
Parati	*Mugil curema*
Mullet	*Mugil platanus*
Guaivira	Various species (5) *
Guaru	*Poecilia viripara*
Enchova	*Pomatomus saltatrix*
Ray	Various species (5) *
Pufferfish	Various species (3) *
Needle	Various species (3) *

* There are several species catalogued; the number beside it refers to the number of species catalogued.

Salinity, rainfall, water temperature, pH, feeding habits, forms of reproduction, etc. will define which species are preferably marine, freshwater, estuarine, marine/estuarine and freshwater/estuarine.

3.5 Tourism Characterization Chart

Table of the Tourism Characterization of the Municipality of Bertioga Survey of Strategies, Main Aspects and Themes, Strengths, Threats, Opportunities, Programs and Actions.				
Aspects and themes	**Strengths**	**Threats**	**Opportunities**	**Programs and Actions**
1. ENVIRONMENT				
1.1 **Sea**	**1.1** Design of artificial reefs for diving and fishing.	**1.1** Overfishing by trawls; pollution from untreated sewage; oil spills from fishing boats; dumping of garbage and waste at sea.	**1.1** Increasing the supply of fish; marine nautical tourism; partnerships with universities for research into marine resources and sporting events.	**1.1** Expansion of the artificial reef project, aimed at curbing predatory fishing. Agreements with universities in the region. Discipline and Regulation of Nautical Tourism. Calendar of sporting events (sailing

				championships, regattas, sailing competitions, etc.)
				swimming, sport fishing, diving, rowing, jet skies, speedboats, etc.). Creation of the Museum of the Sea.
1.2 **Islands**	**1.2** Nests (bird breeding).	**1.2** Disappearance of bird species; detour of natural migratory routes; destruction of vegetation and landscape.	**1.2** Contemplative tourism on the islands and monitored diving in the surrounding area.	**1.2** Municipal zoning with protection actions (monitoring, signposting and publicizing restrictions on the use of islands and contours).
1.3 **Beaches and Dunes**	**1.3** 33 km of beaches. All with good bathing conditions and no potholes. Spearfishing.	**1.3** Pollution of the beach by sewage, dirt and debris on the sand due to too many bathers.	**1.3** Beach sports; strengthening the image of the municipality; sport fishing.	**1.3** Regulation of sports on the sandy beaches and animal access to the beach. Increasing the number of collection points for bathing water laboratory samples. Pollution control. Waste garbage cans along the beaches and a permanent program for cleaning, conserving and maintaining the waterfront gardens.
1.4 Costoes	**1.4** Shore fishing; plenty of shellfish and oysters.	**1.4** Predatory extraction; contamination of crustaceans by pollution; imbalance of the	**1.4** Expansion of the shellfish and oyster trade.	**1.4** Coastal monitoring and preventive signaling.
		food chain.		
1.5 **Jundu**	**1.5** Vegetation to	**1.5** Reduction in the amount	**1.5** Securing sandy	**1.5** Municipal jundu

	protect against sea surges.	of sand; destruction of dunes; invasion of adjacent areas by sea water.	beaches.	protection project.
1.6 **Restingas**	**1.6** Vegetation with a rich diversity of birds.	**1.6** Destruction of several species of birds and imbalance of fauna.	**1.6** Cultivation of native species, production of seeds and seedlings.	**1.6** Implementation of the Municipal Nursery (production of seeds and seedlings).
1.7 **Bertioga Canal**	**1.7** Tourist navigation; great fishing potential (mullet, sea bass, etc.); natural spots for moorings, boat garages and marinas; landscape composition.	**1.7** Compromised navigation; siltation due to excessive sediment; reduced fishing potential; unfeasibility of nautical equipment; accidents due to speeding boats.	**1.7** Improved navigation; development of the marine sector, with the manufacture of small boats and maintenance of marine engines and equipment; more jobs and an alternative for the municipality's economic development. Inter-municipal routes using the Bertioga Canal.	**1.7** Complete and detailed mapping of the banks of the Bertioga Canal, zoning of use and occupation. Regulation of navigation in the Bertioga Channel with beaconing (signaling, buoys, etc.); development of the points that cause strandings.
1.8 **Mango zais.**	**1.8** Ecosystem of Permanent preservation and breeding ground for mullets, shrimps, etc. Habitat for some birds.	**1.8** Landfill of mangroves caused by invasions, slum settlements and environmental imbalance.	**1.8** Research on ecosystem and guarantee estuarine repopulation.	**1.8** Demarcation of Permanent preservation areas (mangroves), installation of fences at points vulnerable to invasion and constant monitoring by land, by river (boats) and by periodic flights. Awareness-raising program for communities living in areas close to the mangroves. Making some parts of the

				mangroves available for research and building walkways with suitable material, providing access for groups monitored for environmental education and researchers.
1.9 **Rivers**	**1.9** Fishy and navigable rivers. Practicing water sports.	**1.9** Pollution due to the discharge of untreated sewage effluent; sand removal in areas that are not permitted; sloping hillsides with an impact on fauna and flora. Loss of sediment and siltation, etc.	**1.9** Development of nautical and sports tourism; artisanal fishing; and sites for the development of aquaculture.	**1.9** Permanent inspection team, zoning to identify riverside areas subject to special projects.
		Compromised water availability.		
1.10 **Atlantic Rainforest, Serra do Mar:** **escarpments** **and slopes.**	**1.10** Rich biodiversity; significant landscape value; habitat for large mammals.	**1.10** Logging of large trees; hunting of wild animals; environmental imbalance; impact on the landscape; landslides and others.	**1.10** Research, Environmental Education and Lodges.	**1.10** Sustainable forest management projects. Regulations for setting up lodges.
1.11 **Springs and** **Water** **Management.**	**1.11** Plenty of drinking water.	**1.11** Reduction of water sources to supply the population.	**1.11** Guaranteed supply of fresh water for future tourist demands.	**1.11** Physical and territorial protection of the springs used to collect water. Campaigns to raise awareness and rationalize water use.
1.12 **Archaeological** **sites.**	**1.12** Sambaquis, Armaçao das Baleias de	**1.12** Natural deterioration of the ruins with the destruction of historical	**1.12** Valuing History, Culture and the Environment.	**1.12** Survey of the current situation of archaeological sites;

	Bertioga; Ruins of Fort Sao Felipe and remnants of quilombos and other sites.	and cultural references.	Development of new attractions.	prospecting plans and programs, with the participation of research institutes. Preparation of specific documents for each archaeological site, describing its history, photos, data, information, restrictions, etc.
1.13 **Indian village.**	**1.13** Demarcated Federal Reserve (FUNAI) of the Tupi-Guarani Indians.	**1.13** Loss of Sustainability and cultural characterization of the community.	**1.13** Handicraft production and sustainable management of natural forest products (palm hearts, orchids, bromeliads, wild honey, etc.).	**1.13** Permanent support for indigenous education; installation of a Craft Center near the village and support for sustainable activities.
1.14 **Trails and waterfalls**	**1.14** Ecotourism, adventure sports, diving wells, swimming and environmental education.	**1.14** Environmental degradation; reduction of ecotourism potential.	**1.14** Expansion of routes and job offers.	**1.14** Mapping the waterfalls with their access points. Discipline the use of trails (supervision, signposting, information and protection) Use of trails for environmental education.

Aspects and Themes	Strengths	Threats	Opportunities	Programs and actions
2. GENERAL STRUCTURE				
2.1 **Access, System Road and Transportation.**	**2.1** Easy access via the Rio-Santos and Mogi-Bertioga highways e Anchieta Imigrantes system.	**2.1** Congestion; excessive accidents; touristic excursions in precarious conditions and truck traffic of load.	**2.1** Increased demand for quality tourism; improvement in the Intercity Transportation System.	**2.1** Disciplining "day tourism". Inspection and control of tour buses and vans. Signposting of the roadside along the municipality. Action to

				control truck traffic (disciplining tonnage, load,
				timetables, etc). Construction of a Bus Terminal.
2.2 **Education**	**2.2** Local history and culture; satisfactory municipal network of public schools.	**2.2** Students and teachers with no knowledge of the city's history; too many teachers from neighboring municipalities, with no commitment to the local community.	**2.2** Environmental and tourism education should be intensified.	**2.2** Implementation of a Municipal Teacher Training Program, including Environmental and Tourism Education. Development, production and publishing of books, leaflets, magazines, newspapers, etc. on the history of the city and its resources and attractions.
2.3 **Health**	**2.3** The city has a treated water distribution network and is expanding its sewage collection network with treatment plants. The city has reasonable hospital facilities and a 24-hour emergency room (including ambulances).	**2.3** The construction schedule is not compatible with the increase in demand, especially during vacations, vacations and the summer season.	**2.3** Attracting public and private investment to expand the health care sector (new hospitals, medical and dental clinics, etc.).	**2.3** Decentralize emergency services, creating other emergency rooms in areas where demand justifies it. Resize hospital equipment and medical and dental resources for vacation and summer season care. Increase the number of ambulances.
2.4 **Security**	**2.4** A town with a low crime rate. Controllable access	**2.4** Process of slumification and land invasions public and private	**2.4** The municipality is easy to fit into the Governmental	**2.4** Controlling the verticalization of buildings on the waterfront beach, with height limits,

Aspects and Themes	Strengths	Threats	Opportunities	Programs and actions
	to the town (only four accesses by land). Low vertical building heights (height restrictions).	attracting bandits, thieves and drug dealers. Clearance for the verticalization of the waterfront; an increase in the number of firefighters to combat fires and other events, with investments in new firefighting equipment, especially for the tallest buildings.	housing policies are preventing slums. The municipality can become a benchmark for urban and beachfront occupation (keeping the height of buildings limited) and guaranteeing waterfront space for leisure, sports and entertainment.	setbacks, occupancy rates and land use. Implementation of a network of fire hydrants and improvements to the Fire Brigade, with new fire engines and other technologies. Increasing the number of police officers living in the city.
2.5 **Economy**	**2.5** Availability of large areas for new productive ventures. Proximity to large urban consumer centers; proximity to the Port of Santos, Cumbica International Airport (Guarulhos) and Congonhas SP Airport. Great fishing and nautical potential.	**2.5** Real estate speculation; disregard for urban and environmental zoning and lack of control, analysis and inspection of developments.	**2.5** Development of new productive activities; business condominiums with new technologies; implementation of new aquaculture and forestry projects; biotechnology, nautical tourism and fishing ventures.	**2.5** Definition of urban zoning, identifying the areas permitted for the projects mentioned in the opportunities. Implementation of a Professional Technical School for the Fishing and Marine sector and for the training of tourism technicians, including labor training for the sector.

Aspects and Themes	Strengths	Threats	Opportunities	Programs and actions
3. INFRASTRUCTURE				
3.1 **Tourist offers**	**3.1** Silveira River	**3.1** Lack of	**3.1** Making the most of	**3.1** Identify, recover and

(natural and cultural attractions).	Indian Village, Sao Joao Fortress (National Heritage 1547). Ruins of Fort Sao Felipe (1553), Itatinga Hydroelectric Power Plant and Village (1910), Bertioga Whale Trap (17th and 18th century). SESC Summer Camp (the largest in Brazil), Mullet Festival, Shrimp Festival, Riviera de Sao Lourenço and the environmental potential already listed in the environmental aspects and themes.	conservation, maintenance and restoration of historical heritage. Lack of awareness on the part of the community and tourists regarding the preservation and protection of equipment. De-characterization of local culture.	the cultural and environmental heritage. Integration of the SESC Summer Camp with the city and use of the excellent social, cultural, sports and leisure facilities at the Summer Camp.	publicize all the artistic and cultural manifestations of the municipality. Creation of a Cultural Center, registration of artists, craftsmen, poets, writers, musicians, painters, sculptors, etc. A program to publicize the municipality's cultural potential in municipal schools. Promoting cultural events and meetings. Elaboration of the Festival Calendar, with greater community involvement and participation. Publicizing the calendar in the media in general.
3.2 Offer of Means of Accommodation and Food. Entertainment, leisure and services.	3.2 Good network accommodation (hotels, hostels and camping). Offering properties for rent.	3.2 Unbalance-supply in relation to demand.	3.2 Attracting new people lodging, food, leisure, entertainment and service businesses. Increased generation of income and jobs.	3.2 Creating incentives and tax incentives for businesses that generate income and jobs. Create training and improvement courses for employees of existing accommodation establishments.
3.3 Specific	3.3 Public policies to	3.3 Political and party	3.3 Creating incentives	3.3 Contracting and

legislation.	encourage tourism. Enhancement of the Municipal Tourism Council and strengthening of the Municipal Tourism Fund. Improving and streamlining the Municipal Tourism Office.	interference in the Council. Lack of specific policies for tourism development. Lack of budget allocation for basic demands.	for tourism. Preparation of the Tourism Master Plan.	Implementation of a Municipal Tourism Plan.

| legislation. | encourage tourism. Enhancement of the Municipal Tourism Council and strengthening of the Municipal Tourism Fund. | interference in the Council. Lack of specific policies for tourism development. Lack of budget allocation for basic demands. | for tourism. Preparation of the Tourism Master Plan. | Implementation of a Municipal Tourism Plan. |

CHAPTER 4 - CHARACTERIZATION OF ENTERPRISE

4.1 History and Presentation

The Itatinga Hydroelectric Power Station[8] is located at the foot of the Serra do Mar in the municipality of Bertioga (SP). The area belongs to Companhia Docas do Estado de Sao Paulo - CODESP (the company that manages the Port of Santos - SP) and its construction began in 1906.

It was inaugurated on October 10, 1910 and built to generate electricity for the Port of Santos, with the aim of electrically driving equipment, machines and motors and lighting the areas of the quay, warehouses, offices, etc. which until then had been powered or lit by steam engines.

Its conception began in 1890, when the British drew up studies and projects for the construction of a hydroelectric power station to generate energy for the Port of Santos.

In 1903, the Companhia Docas de Santos, managed by the Guinle and Gaffrée families, who were responsible for the concession of the port, acquired part of the **former Fazenda Peales** at the foot of the Serra do Mar. The area is located in Bertioga, formerly the District of Santos, now the municipality of Bertioga, whose political and administrative emancipation took place on May 19, 1991.

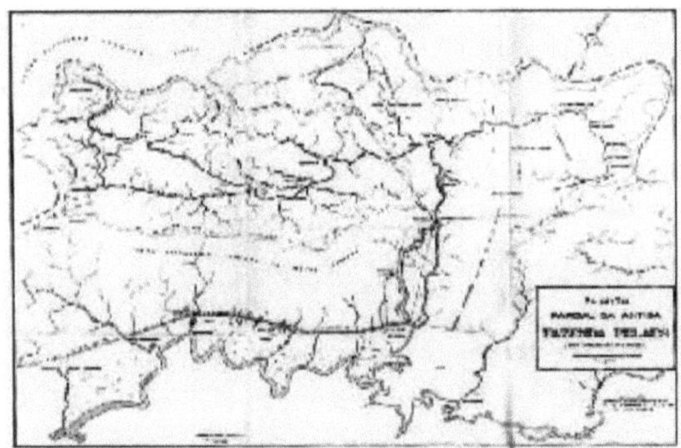

Map of the former Fazenda Pelaes (CODESP Collection)

The area chosen for the construction of the plant, 30 km from the Port of Santos, aimed to take advantage of the Itatinga River, whose sources are on the plateau, and allowed for the construction of a catchment at a height of 765 meters above sea level, in a place where there

---

8 **ITATINGA:** The etymological origin and meaning of the word "Itatinga": it is a term of Tupi origin, corrupted from "*Ita-tinga*" = white stone.

was originally a natural waterfall. Water chambers were built and five pipes run down the mountain to the plant's five generators.

The plant was designed by engineer Guilherme Benjamim Weinschenck, and construction began in 1906 and was completed four years later. (Historic photo with the staff on site).

Historical photo of the site managers at the time of Itatinga's construction (CODESP Collection)

At the time, it was one of the largest run-of-river hydroelectric plants in the country, and also became the one with the highest waterfall. The unit has an output of 15,000 kilowatts (obtained from the Itatinga River's flow of 3.3 m^3 /s). Since its inauguration, the plant's capacity has enabled the company to maintain the operation of its port facilities, which have an average demand of 10,000 kilowatts, and to sell the leftovers to the region's energy supply concessionaires.

Until 1927, the plant supplied electricity to the municipalities of Santos, São Vicente and neighboring towns; it even fueled the construction of the Henry Borden Plant in Cubatâo-SP.

In 1925, during the capital's energy crisis, Itatinga also supplied five thousand kilowatts to São Paulo.

Along with the installation of the power plant and the electricity generation system, a village was built, designed to meet all the needs of its residents, such as a cinema, school, chapel, bakery, sports facilities and others.

Image of the crossing. Boat used to cross the river (IPECAB Collection)

Mooring for boarding the speedboat across the Itapanhaù River (IPECAB Collection)

Access between the plant and the urban area of Bertioga is by boat across the Itapanhaù River. After crossing, the boat reaches a "Portinho", where there is a platform for boarding an electric train (or cable car), a means of transportation for residents and visitors to Vila de Itatinga.

The 7 km long railway, although currently electrified, still has a steam locomotive built in 1909.

Bondinho Station (CODESP Collection).

Photo of the 1909 steam locomotive (CODESP Collection).

The Itatinga Hydroelectric Power Station and its village are located in an area adjacent to the Serra do Mar State Park, considered one of the main Conservation Units in the State of São Paulo (created by Decree 10.251, of 30/08/1977, as amended by Decree 13.313, of 06/03/1979).

The entire area where Vila Itatinga is located belonged to the old Fazenda Pelaes, a productive property from the days of colonial Brazil. The site was home to old families who lived off banana cultivation. Even today it is possible to find intact stretches of "trolleybus", pulled on the iron rails by animal traction.

Nearby there are remnants of ancient quilombos, pirate refuges, hermitage ruins, archaeological sites (sambaquis) and other important quotes from the region's leading historians.

When the first Master Plan for the municipality of Bertioga was drawn up (1995/1996), under the coordination of architect and urban planner Nilo Nunes, the area of Itatinga, Pròprio Federal, was already indicated.

(CODESP), as a **Zone of Historic-Cultural Interest (ZIHC).** At the time, there *was already* some thought of guaranteeing some protection and preservation through the Municipal Land Use and Occupation Legislation and the Master Plan Law itself.

With the withdrawal of the Master Plan Project from the City Council by the subsequent Municipal Administration, with the claim that some changes would be made, the expectation was that if no progress was made, at least the original proposal would be maintained.

Although the first project was fully used as the basis for the other proposal, there was a relaxation in the restrictions on the use and occupation of the ZIHC. The original proposal, which defined Itatinga as a Zone of Historical-Cultural Interest in the municipality of Bertioga, was left vulnerable in the sense of protecting the historical heritage in question.

As a result, the entire structure of Itatinga (comprising the water catchment at the 765-meter level of the Serra do Mar, pipelines, the powerhouse building, winches, rails, houses in the village, cinema, school, bakery, medical center, equipment, Our Lady of Conception Chapel, sidewalks, boulevards, electric cable car, port, railway, etc...) was left unprotected.

After that, CODESP publicly announced that the company didn't have the resources to "modernize" the hydroelectric plant and presented the solution of outsourcing the services (privatizing the plant). CODESP claimed that in the hands of the private sector, the plant would "increase its power output, increase the supply of energy, generating new jobs and more taxes for the municipalities in the region".

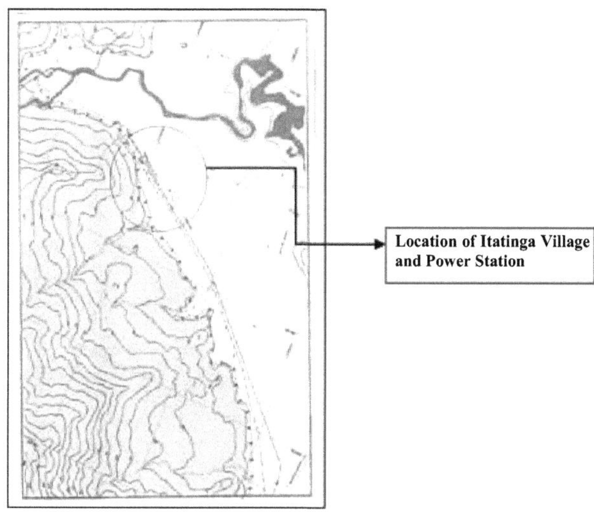

Altimetry of the Plant's Location

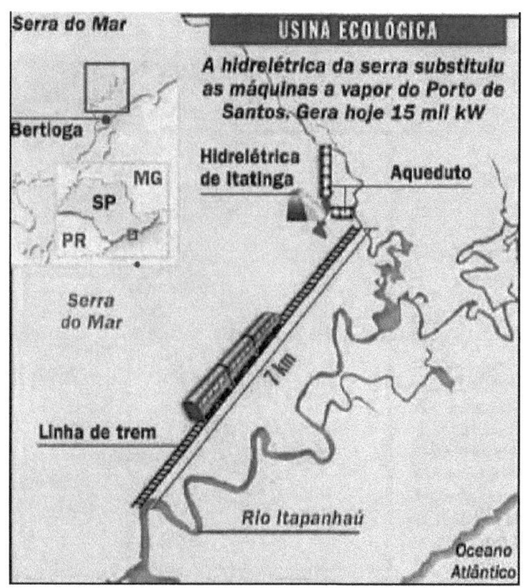

See the following clipping from the Santos newspaper "A Tribuna": *"CODESP announces the outsourcing of its plant in Itatinga" (February 6, 1999)*.

Privatização

Codesp anuncia a terceirização da usina que mantém em Itatinga

O negócio será realizado ainda neste semestre, segundo a direção da empresa

Da Sucursal

A Usina Hidrelétrica de Itatinga, em Bertioga, que abastece todo o Porto de Santos, deverá ser terceirizada ainda neste primeiro semestre. O anúncio foi feito ao prefeito Luiz Carlos Rachid, recentemente, pelo diretor presidente da Companhia Docas do Estado de São Paulo (Codesp), Paulo Fernandes do Carmo, durante audiência na sede da estatal, em Santos. Rachid disse que a privatização vai ajudar no desenvolvimento econômico do Município.

O encontro teve também a participação do secretário de Administração, Finanças e Jurídico do Município, Antônio José Fabris, e do diretor de Turismo, José Carlos Vasques Rodrigues. Paulo Fernandes garantiu ao prefeito que o processo de terceirização da usina é simples.

O primeiro passo será a publicação de edital abrindo a concorrência pública para as empresas interessadas. Em seguida, a Codesp promoverá uma audiência pública em Itatinga, para que os municípes se pronunciem se querem ou não a terceirização. A audiência, segundo Paulo Fernandes, deverá acontecer em abril.

Rachid enfatizou que a transferência da usina à iniciativa privada possibilitará também a geração de novos empregos e impulsionará a economia local, visto que a empresa vencedora vai contribuir com a economia do Município. "Vamos trabalhar com a Codesp para que a audiência pública seja um sucesso", afirmou o prefeito.

O secretário de Finanças, Antônio Fabris, lembrou que a partir da terceirização a empresa que ganhar a concorrência terá inscrição estadual no Município. Hoje, segundo ele, o endereço da Usina de Itatinga, localizada em Bertioga, tem a Avenida Rodrigues Alves, em Santos, onde se localiza a sede da Codesp. "A medida só trará ganhos para Bertioga", destacou.

Parceria — O prefeito Rachid aproveitou a audiência para reforçar o pedido já registrado em ofício à Codesp, propondo parceria com a estatal na implantação do sistema de iluminação pública na Rua Manoel Gajo, trecho compreendido entre o portinho de Itatinga e o primeiro poste de iluminação da Electro, numa distância de aproximadamente 400 metros, nos moldes do sistema utilizado na Vila de Itatinga.

Outro ofício solicita também parceria com a Codesp para que a empresa ceda caminhões destinados ao transporte de material utilizado no nivelamento da Rua Manoel Gajo, entre a Rodovia Rio-Santos e o portinho de Itatinga, principal acesso àquela vila.

According to CODESP itself, over the last few years, the company has not carried out any work or purchased any equipment to "modernize" the plant, but has only carried out operational maintenance.

Paradoxically, this situation ended up guaranteeing

the Itatinga Power Station and its entire structure and collection a degree of preservation that has maintained its original characteristics since its inauguration on October 10, 1910.

In addition, while praising the zeal and competence of this maintenance, it was possible to secure technological rarities. Among the original equipment still in full working order are the five hydraulic turbines with a 2.14-meter wheel and 16 metal shells built in 1906 by J.M. Voith of Heidnheim (Germany), three-phase electric generators from General Electric (USA) from the same period, and a huge range of other small equipment and installations.

All of this historical, architectural, technological and cultural heritage is located in a natural setting with immense environmental and landscape wealth, surrounded by the Atlantic Rainforest on the slopes of the Serra do Mar.

In the small village, where there are around 70 houses along the only street, you'll find countless matrices of native palm trees, a variety of birds and butterflies, flocks of green parakeets, swallows, sabiàs, sairas, teiùs, parrots and even toucans, as well as various species of insects, beetles, ants and wasps.

Photo of Itatinga Village, 2013 (Source: Geribello, 2016)

The verdant Serra do Mar forms a "backdrop" to this wonderful scenery and offers a variety of trails that run along the mountain slopes and rivers.

Image of the Chapel and its surroundings (IPECAB Collection)

Steam locomotive "Lavoura" No. 01 (IPECAB Collection)

The region's rivers and streams originate on the slopes of the Serra do Mar, typical of the Atlantic Rainforest. These rivers are bordered by dense vegetation and have cold, crystal-clear and clean waters, to the point where it is possible to see pools and natural pools with many schools of colorful fish.

Meanders of the Itapanhaù River (IPECAB Collection)

On the banks, the large number of pebbles shows the force of the water that crosses the area during heavy rainfall; and, looking at the vegetation, we find a large number of bromeliads, orchids and other epiphytes.

Any visitors walking along the trails in the area will find footprints and traces of small mammals, reptiles and other wildlife.

Image of a heron on a boat on the Itapanhaù River (IPECAB Collection)

4.2 Tombstone

Listing means, in practical terms, the formal recognition by the public authorities of the cultural value of an asset. The cultural heritage of a community, a city, a people or a country is the set of monuments, buildings, documents, sites, landscapes, knowledge, celebrations, forms of expression, and many other elements considered important for their aesthetic, emotional, environmental, artistic, historical, technological, etc. values.

Landmarking represents the inscription and registration of an asset in a special inventory book called the "Book of the Tomb", which symbolically recognizes its cultural value and social interest.

When the public authorities consolidate the **listing**, they become the main agents in the process of defending, conserving and eventually restoring the listed heritage.

If the **landmark** is on private property, this does not render the area unusable; on the contrary, it allows it to be used for specific purposes, as long as the original characteristics of the landmark are not altered.

The notion of monument or building, architectural or urbanistic ensemble, includes not only the isolated creation, but also the environment in which it is inserted. The monument, the building or any architectural ensemble is recognized as having cultural value as a result of time and humanistic significance.

The concern about the possible "privatization" and "modernization" of the plant and its facilities leads to the question of whether there will be a replacement of equipment; whether the "modernization" of the energy generation process will guarantee the preservation of the historical, architectural, cultural, environmental, landscape and tourist value of the plant, its facilities in the village and its equipment and its surroundings.

Is it possible to reconcile a "modernization" of the Electricity Generation System without impacting the environment and without de-characterizing this rich heritage?

And will the cable car, the "portinho", the railway line, the houses, etc., be kept? For all of these reasons, the **landmark** was proposed.

On 21.02.2000, the Instituto de Pesquisa e Ciências Ambientais de Bertioga - IPECAB (Bertioga Institute of Research and Environmental Sciences), an NGO (Non-Governmental Organization), sent a letter to the President of CONDEPHAAT, explaining the reasons for opening the Listing Process, attaching documents, available plans, photographic surveys, publications, etc., requesting the opening of the relevant Listing Process.

The letter was formally sent via SEDEX mail, post number SE-747072066, and received in Sao Paulo - Capital, the following day, February 23, 2000, according to the acknowledgement of receipt.

On July 11, 2000, the Council for the Defense of the Historical, Archaeological, Artistic and Tourist Heritage of the State of São Paulo - CONDEPHAAT published in the Official Gazette of the State of São Paulo, Section I, page 34, the **"Official Notification"** of the Opening of the Listing Process for the Itatinga Power Plant and all the annexed facilities in the municipality of Bertioga **(CONDEPHAAT Listing Process No. 40036/2000).**

A community that preserves its history, its past and its memory will be permanently in tune and prepared to achieve sustainable development.

It should not be overlooked that the use of a "listed property" as a cultural, social and leisure space can induce and provoke a qualitative transformation. Therefore, the revitalization of the Itatinga area, with the implementation of an **Ecotourism Theme Park**, can leverage a process of local and regional tourism development.

CULTURA

CONSELHO DE DEFESA DO PATRIMÔNIO HISTÓRICO, ARQUEOLÓGICO, ARTÍSTICO E TURÍSTICO DO ESTADO DE SÃO PAULO - CONDEPHAAT

Notificação

De acordo com o que dispõe o artigo 142 do Decreto 13.426, de 16.03.79, notificamos a todos os interessados que a Presidência do Conselho de Defesa do Patrimônio Histórico, Arqueológico, Artístico e Turístico do Estado - CONDEPHAAT, com base no item IV do artigo 169 do Decreto Estadual 20.955/83, avocou a decisão e aprovou a abertura do processo de estudo de tombamento da Usina de Itatinga, bem como de todas as instalações anexas, localizadas no Município de Bertioga.

Nos termos do parágrafo único do já citado artigo 142 e do artigo 146 do mesmo Decreto, a deliberação de abertura do processo de tombamento assegura, desde logo, a preservação do bem até decisão final do autoridade competente, ficando, portanto, proibida qualquer intervenção que possa vir a descaracterizar a referida área, sem prévia autorização do CONDEPHAAT, além de poder ser punido o descumprimento do acima disposto com as sanções penais previstas no artigo 166 do Código Penal Brasileiro e da Lei 7347, de 17.07.1985.

(8-11-12)

Publication in the Official Gazette of the State of São Paulo

4.3   Project Concept

4.3.1 Basic Concepts

The proposal to set up an *Ecotourism Theme Park at the Itatinga Hydroelectric Power Station* aims to guarantee the preservation of the environment and the historical, cultural, architectural, landscape, technological and scenic heritage through tourist use.

With the CONDEPHAAT listing process, it became essential to formulate alternatives for the use, enhancement and utilization of this historic site.

The Itatinga Ecotourism Theme Park was conceived according to the ecological concept of proposing and **encouraging human relationships with the environment** and provoking a healthy and sustainable interaction with ecosystems.

It is a proposal that guarantees the spread of the conservation movement and the confirmation that the conservation of natural resources is not incompatible with economic and social development.

The park is part of the "Family Park" concept, which aims to provide leisure for the whole family in the same environment.

4.2.1 Commitment to Environmental Education

The Park was conceived to offer much more than amusement, leisure, recreation and entertainment products.

The planning aimed to exercise environmental education in every detail, trying at all times to awaken a new awareness in visitors.

Environmental education:

The conception of the Itatinga Ecotourism Park is geared towards a holistic vision and proposes a model of harmonious relationship with nature. It adopts proposals and attitudes of integration and participation, where each visitor is encouraged to fully exercise their **sensory**[9] and **perceptive**[10] capacity.

The aim was to explore the richness of the ecosystems involved, not only in order to develop an understanding of the complex environmental systems of which we are a part and to disseminate information, but also to inspire everyone to engage in building a world that is supportive, ethical, fair, happy and socially sustainable.

4.2.2 Methodology and Preliminary Zoning

The park as a whole was designed to take advantage of all the potential of the area, **sizing up the support capacity** and existing structure of the village of Itatinga, including access roads,

---

9 **SENSORY:** Pertaining to the brain or the sensory (suitable for transmitting sensations) relating to sensation, the senses.
10 **PERCEPTIVE:** Pertaining to perception. Faculty of perceiving. Not just "looking", but "feeling".

means of transport, buildings, spaces and existing equipment.

The Methodology for Analyzing Potential and Resources started with a global view of the area, immediately establishing a **preliminary zoning**, considering **the total** area of **the Park to be** the sum of all the areas of the Itatinga Hydroelectric Power Plant, according to the CODESP General Plan, reference I-VII-11188, **totaling 16.327,070 m²** (described in Table 1 - Areas), although the area actually in **intensive use** (area to be used directly by visitors) and the **special use area** (areas needed for Administration, Maintenance, Workshops and Services) amount to **320,750 m²** , i.e. only around 2% (two percent) of the total area.

The guidelines of the Regulations for Brazilian National Parks, according to Federal Decree No. 84.017 of September 21, 1979, suggest that *the* Management Plans: *"make the preservation of protected ecosystems compatible with the use of the benefits and that the guidelines guarantee adequate ecological management"*.

The conception of the **Itatinga Thematic Park,** in line with the guidelines for Brazilian National Parks, proposes a **zoning of** the **total area**, guaranteeing an **intangible zone** (that which is primitive, intact, without human alterations, with a high degree of preservation and integral protection of ecosystems and genetic resources), **primitive zone** (with areas that facilitate scientific research and environmental education) and **extensive use zone** (natural areas, with low human impact), although it does have some **trails** and offers access to the **Itatinga River and the proposed viewpoints**, an area where some **interpretive programs** will be carried out.

The development was designed, zoned and sectored, inspired by the rules that define and characterize National Parks (Federal Decree No. 84.017 of September 21, 1979).

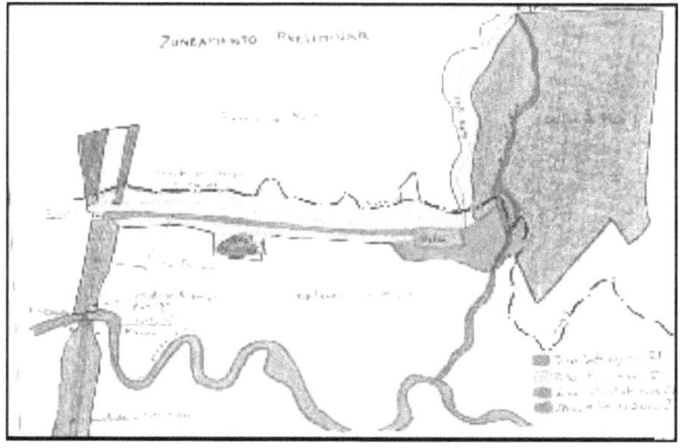

Sketch of the Park's Preliminary Zoning (authored by Nilo Nunes)

A significant part of the Total Area of the Itatinga Hydroelectric Power Plant, endowed with exceptional natural attributes, is part of the **Serra do Mar State Park**, an object of permanent preservation.

Serra do Mar State Park: The limits in the municipality of Bertioga are defined, to the south, on the border between Bertioga and Santos, by the altimetric level of 100 meters to the source of the Jaguareguava River, following the course of the river to a level of 20 meters and continuing at this level to the limit with the municipality of Sao Sebastiao in Boracéia. (Decree No. 149, of 15/08/1969; State Decree No. 10.251, of 30/08/1977, amended by Decrees 13.313 of 06/03/1979 and 13426 of 16/03/1979).

In order to make the preservation of protected ecosystems compatible, zoning was determined according to the characterization proposed in Article 7 of the aforementioned Decree.

<u>ZONES</u>[11] : - **I INTANGIBLE ZONE (IZ)** - this is the primitive, intact **zone**, with a high degree of preservation, with no alterations tolerated. It functions as a repopulation matrix. It is dedicated to the integral protection of ecosystems, genetic resources and environmental monitoring.

Objective: Preservation guaranteeing natural evolution.

II PRIMITIVE ZONE (ZP) - is a zone where little or no human intervention has taken place, containing species of flora and fauna or natural phenomena of great scientific value. Characteristics of a transition zone.

Objective: To preserve the natural environment and facilitate scientific research, environmental education and primitive forms of recreation.

III EXTENSIVE USE ZONE (EUZ) - made up of natural areas, which may show some anthropogenic alteration; it is characterized as a transition zone between the primitive zone and the intensive use zone.

Objective: Preservation of the natural environment and activities with minimal impact, while offering access and facilities for educational and recreational purposes.

IV - INTENSIVE USE ZONE (ZUI) - natural areas or those altered by man. The environment is kept as close to natural as possible and should contain a visitor center,

---

11 **ZONE:** (1) Belt, strip. (2) A region characterized by certain peculiarities, generally delimited by external appearance, the nature of activities, land use, environmental restrictions, etc.

museums, other facilities and services.

Objective: To facilitate intensive recreation, environmental education and ecotourism.

SECTORS[12] **: MAIN ACCESS SECTOR** - from the park's main gate to the "Bondinho" station.

SECTOR OF ROUTES AND THE RAILROAD - from the station to the town.

TRAILS AND MIRRANTS SECTOR - includes all the areas around the village.

VISITOR SECTOR - Itatinga village area.

AREAS :[13]

MAIN ACCESS SECTOR

(1) Access to the Portal
(2) Portal.
(3) Reception.
(4) Parking.
(5) The area between the Portal and the Mooring.
(6) Mooring area.

ROUTE AND RAILROAD SECTOR

(1) Area of the Itapanhaù River Crossing Strip.
(2) Bondinho Station area.
(3) Station Port area.
(4) Area of the railway line right-of-way.

TRAILS AND VIEWPOINTS SECTOR

(1) Archaeological site area
(2) Contemplation Area.
(3) Trail areas.
(4) Lookout areas.

VISITOR SECTOR

(1) Administration, Maintenance, Workshops and Services Area.
(2) Village area.
(3) Rest and eating area.
(4) Sports Area

4.4 Parks in the world and in Brazil

---

12 **SECTOR:** (1) Subdivision of a zone, a district, a section, etc.
13 **AREA:** Portion of the Sector or segment of the sectorization with specific purposes and uses.

4.4.1 International References

According to data from the WWA - World Water Parks Association, the establishment of parks has developed very rapidly over the last thirty years, mainly in the United States, Europe and Asia. Today, it is one of the strongest economic activities in the world, generating US$ 17 billion annually, with a growth rate of around 12% a year. The United States and Canada alone account for more than half of this figure, US$8 billion, with 300 million people visiting the country every year.

Leisure and tourism are responding to the growing need for new entertainment options outside the home, especially for increasingly urban populations.

Parks are divided into ***outdoor*** (open environments with limited space) and ***indoor*** (closed environments with limited space). They can be **dry** or **aquatic**. They can have various amusements, toys and attractions with specific themes, in other words, they can have **themed attractions**.

The **Itatinga Ecotourism Theme Park** would fit in as an **Outdoor Park** and could be dry and aquatic at the same time, because as well as incorporating all the environmental, landscape, architectural and technological **themes**, etc., it could exploit the quality of the waters of the **Itatinga River**, offering a diverse range of activities and entertainment.

In a survey carried out and published at the beginning of this century by the ***International Association of Amusement Parks and Attractions*** (IAAPA), it was discovered that the 50 main parks in the United States received around 147 million people a year, generating resources in the order of US$ 4.3 billion. Smaller parks contributed more than US$2 billion to the sector's revenue.

These figures are also significant in other markets, such as Asia and the Pacific, where there is a structure of parks that generates resources in the order of US$ 5 billion, with an approximate flow of 115 million people/year, and in Europe, where the resources generated are approximately US$ 3.0 billion, for an approximate flow of 75 million people/year.

Some examples:

	Parks/Types	Name	Location
01	Outdoor/ Dry Fun Parks	Playcenter	Brazil - São Paulo
		Six Flags	California - USA

No.	Type	Park	Location
02	Themed Outdoor/Dry Parks	Beto Carrero World	Brazil - Santa Catarina
		Euro Disney	France - Paris
03	Outdoor/Aquatic Amusement Parks	Wet'n Wild	Brazil - Salvador
		Water World USA	USA - Sacramento- CA
04	Themed Outdoor/Aquatic Parks	Beach Park	Brazil - Fortaleza - CE
		Blizzard Beach	USA - Florida
05	Indoor/Dry Fun Parks	Playland	Brazil - São Paulo
		Games Works	USA - Seattle
06	Themed Indoor/Dry Parks	Monica's Park	Brazil - São Paulo
		Liserberg	Sweden - Stockholm
07	Indoor/Aquatic Fun Park	Bakken	Denmark - Copenhagen
08	Themed Indoor/Aquatic Park	Duinrell	Netherlands - Wassenaar

4.4.2 Parks in Brazil

The amusement market, and especially the theme park market in Brazil, is in a growth phase. It's a young sector and still not very numerous, but in recent years it has shown a capacity for accelerated growth.

Pioneering projects such as Playcenter in Sao Paulo, Beach Park in Fortaleza and Beto Carrero World in Santa Catarina are now being joined by several other theme and water park projects throughout Brazil.

Theme parks belong to the service sector and are capable of redistributing income, attracting foreign currency, generating direct and indirect jobs, boosting other economic sectors, increasing tax revenue, promoting regional development and inducing new investments with social benefits.

Theme Parks are enterprises based on differentiated THEMES in the setting of their attractions.

Brazil has 195 theme parks, water parks, amusement parks and even mobile parks. According to the Association of Amusement Park Companies of Brazil - ADIBRA, the sector had a turnover of around R$ 430 million in 2000, and the balance sheet for 2001 to 2005 shows an increase in attendance of 25%.

According to research carried out at the end of the last decade by the Development Support Foundation (FADE), the World Tourism Organization (WTO) and EMBRATUR, theme parks have the marketing objective of stimulating tourist activity. Own management prevails in 64% of the enterprises in Brazil and outsourcing is more prevalent in amusement parks.

MAIN CONCLUSIONS
84% favor marketing through direct contact with people (sales at the moment the public enters).
In specific theme parks, (37%) of sales are made through travel agencies and operators.
The best months for the specific theme parks are the high season months: January, February and July.
The large parks in Brazil have a capacity to absorb visitors/day = 17,500 people.
The average annual number of visitors to the theme parks is (33%), rising to (53%) in the high season.
The specific theme parks have the highest demand, (43%) annual demand and (70%) during the high season.
Water parks have the lowest demand (21%).
Amusement Parks: (37%).
The average stay of Theme Park users = 04 (four) hours (the size, the number of attractions and the single ticket charge influence the stay of visitors).
The specific theme parks operate under the Single Entry System.
Amusement Parks (50%) with a single ticket. (50%) with ticket per attraction/toy.
Attractions and services: quick meals in cafeterias, with food and beverage services being outsourced.
Store rental is more common in larger parks.
Average ticket prices: (more expensive for specific theme parks - average = US$ 22.22)

Average Annual Value of Single Tickets to Theme Parks in Brazil (in US$).

TYPES	STAGE			
	HIGH	AVERAGE	LOW	AVERAGE VALUE
Specifics	22,88	22,39	21,39	22,22
Aquatics	18,90	15,59	15,25	16,58
Fun*	17,91	16,91	16,91	18,24

SOURCE: FADE/EMBRATUR survey * Amusement parks charge a single ticket

Average daily visitor spending, by type and by season in Brazil (in US$).

TYPES	STATION			
	HIGH	AVERAGE	LOW	AVERAGE
Specifics	37,31	35,32	33,30	35,31
Aquatics	24,40	22,40	10,94	19,24
Fun*	9,30	10,70	8,95	9,65
Source: FADE/EM survey			BRATUR	

***Direct job creation:**

- An average of 0.02 workers per visitor.

- There are variations in the averages for each type of resort, dictated by the range of attractions and services on offer.

ABSORPTION OF LABOR IN BRAZIL'S THEME PARKS		
TYPES	EMPLOYMENTS - Average per park	EMPLOYEES - visitors
Specifics	550	0,03
Aquatics	64	0,007
Fun	55	0,02
SOURCE: FADE/EMBRATUR		

DISTRIBUTION OF THE WORKFORCE IN BRAZIL'S THEME PARKS BY RIDE OPERATION (IN %) OF TOTAL EMPLOYEES

56% in Amusement Parks

29% in Water Parks

24% in Theme Parks

SOURCE: FADE/EMBRATUR

PAYROLL FOR TOY OPERATIONS	
In amusement parks	50% of salaries paid
In specific theme parks	24% of salaries paid
In Water Parks	20% of salaries paid

GROSS MONTHLY SALARY (AVERAGE)			
For Jobs in Brazil's Parks (in General)		**For Jobs in Brazil's Theme Parks**	
4,13 Minimum wage (R$ 937.00 Jan/17)	R$ 3,776.11 (s.m. Jan/2017)	4.64 Minimum wage (R$ 937.00 Jan/17)	R$ 4,347.68 (s.m. Jan/2017)

COMPOSITION OF REVENUE (Single ticket)
87% of Amusement Parks' turnover
64% of the turnover of Specific Theme Parks
58% of water parks' turnover
SOURCE: FADE/EMBRATUR

COMPOSITION OF FIXED PAYROLL COSTS (SALARIES, SOCIAL OBLIGATIONS AND FIXED EXPENSES)			
FIXED	Amusement Parks	Aquatics	Specific Themes
	86%	57%	60%
SOURCE: FADE/EMBRATUR			

<u>Observations</u>: There is a tendency for newer parks to commit around 70% of their net revenue to financial expenses. Parks that have already consolidated significantly reduce these figures and spend 45% of total revenue on personnel.

<u>Variable costs</u>: Advertising = 30% (on average) of the variable cost (in recently established parks).

Amusement Park = 15% (on average)

Final remarks on the research:

☐　Equipment in general represents 70% of the total (initial) investment.

☐　Profitability: Activity level, measured in average visitor frequency, is equivalent to (33%) of attendance capacity. The break-even point for Theme Parks is in the order of (18%), i.e. you can already achieve non-negative net results. Operating safety margin: (13%).

☐　27%) net profit per year demonstrates efficiency in both revenue generation and cost management. SOURCE: FADE/EMBRATUR

TABLE BELOW: DESCRIPTIVE SUMMARY OF SOME THEMATIC PARKS IN BRAZIL. THE PARKS HAVE JUST BEEN MENTIONED AS EXAMPLES OF THEMATIC PARKS AND THEIR MISCELLANEOUS CHARACTERISTICS:

PARK NAME	DESCRIPTIVE SUMMARY
BETINHO'S FARM CARREIRO	It is located on Av. Marginal Pinheiros, on the corner with Avenida Agua Espraiada, in Sao Paulo (SP). It's a mixture of amusement park, circus and farm. Although it is open to all ages, it is especially aimed at children between 8 and 10 years old. There are baby goats, geese, teals, pigs and horses, as well as turkeys, chickens and a pony carousel. Of the typical amusement park rides, the highlight is the carousel of live ponies, attached to each other with ropes, they go round and round in a small arena. http://www.betinhocarrero.com.br/
NEW MÔNICA PARK	It's a themed amusement park aimed at children up to the age of 13, with rides for boys and girls and all the Monica's Gang characters. It's bigger than the previous one and has many attractions. Among the new attractions are the Astronaut rollercoaster, which simulates a space station in a stellar environment; the themed houses of the main characters (Monica, Magali, Cebolinha and Cascao); Madman's bumper cars; Piteco's climbing (the environment imitates the character's prehistoric setting, with volcanoes and rocks); the princes' and princesses' castle (where children can play fairy tales), and others. In the TV studio, kids can learn how cartoons are put together. In addition to the park, kids can watch plays, 3D movies and concerts. There is also a food court, fridges and stores. Parque da Mônica is

	located at Shopping SP Market - Avenida das Nações Unidas, 22540 - Jurubatuba, Sao Paulo - SP. http://parquedamonica.com.br/
MAGIC PARK APARECIDA	It is a religious, cultural and recreational park. It is next to the Basilica of Our Lady of Aparecida. It has three types of attractions. The first is religious. In the park, the birth of Christ is reenacted every 30 minutes in a nativity scene made up of 84 computerized puppets. There are several themed areas that reproduce different styles (medieval and western) and cultures - Arab, French, Italian, Mexican, Nordic, Portuguese and Roman. Everything imitates the typical architecture of these cultures. It has games
	and 19 options of amusement toys. It also offers the World in Miniature, which shows 98 replicas of historical monuments from 31 countries. It is located on Rodovia Presidente Dutra, exit km 74 in Aparecida (SP). http://www.amadeusturismo.com.br/ Parques/Parques T/Nac/Mag
	ic P.htm
VIVA PARQUE (FORMER GUGU PARK)	Viva Parque (formerly Fantasy Acqua Park) is a good facility, especially when it's hot. Since it began operating in the 1990s, it has undergone several renovations and modernizations. The park has options to please all kinds of people, of all ages. There are extreme water slides and shallow pools for children to play in. There are also trails, a waterfall and areas to relax. Children under 5 and adults over 65 are free. Located in the middle of the Atlantic Forest, in the municipality of Juquitiba, the Viva Parque Aquatic Park is set in an area of 332,000 square meters, of which 90,000 square meters are set aside for leisure activities and the rest is virgin forest, designated as an environmental preservation area; the Aquatic Park is truly a leisure option for those looking for fun in the region. The park is open on Saturdays, Sundays and public holidays from 9am to 5.30pm. Viva Parque's address: BR 116 (Régis Bittencourt) Km 320, SP 57 - Estrada do Jacuba, 2011 - Juquitiba SP - Brazil. http://www.vivaparque.com.br/
HOPI HARI	It opened in 1999. It offers an amusement park with its own language. There are five themed areas. At the entrance is Kaminda Mundi, where the first inhabitants of Hopi Hari arrived. There is also a bumper car, only on water. The 76-hectare park also has 35 attractions. The Hadikali, a hang gliding simulator, Chabum, a water playground from Sesame Street, La Tour Eiffel, a replica of the Parisian Tour Eiffel, with an elevator that descends in free fall from a height of 69.6 meters and Ekatomb, an earthquake simulator. It is located at Rodovia dos Bandeirantes, km

	72.5 in Vinhedo-(SP). It has a full medical clinic, facilities for nursing mothers, baby changing facilities in various parts of the park, access for people with disabilities, parking for 3,500 cars, among other things. The park offers discounts for groups on excursions. http://www.hopihari.com.br
SKI MOUNTAIN PARK	It opened in 1998 and is located in the tourist resort of Sao Roque. It has attractions for all ages and tastes. The park is on one of the city's beautiful mountains at 1200 meters above sea level and with 320,000 m^2 of Atlantic Forest. You can
	"ski" to your heart's content (artificial skiing). It has a 500-meter-long polyethylene slope for teaching the sport. The park has cable cars, trails, mountain climbing, abseiling and a 400-meter-long toboggan run. Other attractions include a ski and snowboard slope, an ecological skating rink, paintball, a playground, a climbing tower and more. There is also a steakhouse, coffee shop, tea house and a sweet shop known as *Delicias da Vovó*. Access to the park is via Estrada da Serrinha, 1500, in the Cambarà district. It is only 54 km from the city of São Paulo, with access via the Castelo Branco and Raposo Tavares highways. http://www.skipark.com.br
WALTERWORLD PARK (Poços de Caldas)	It's a theme park in Poços de Caldas (MG). Poços de Caldas, one of the most visited cities in the south of Minas Gerais, is famous for the medicinal power of its sulphurous waters, the beauty of its Murano crystals and its intense cultural program. The Walter World park covers 150,000 square meters and is part of the *Thermas Hotel Walter World* complex. The attractions are aimed at children, especially those aged between 3 and 8, but parents can also enjoy themselves in some places. There are amusement rides such as the Children's Carousel, Ferris Wheel, scenes reminiscent of children's stories, such as ghosts, Snow White and the Seven Year Olds, Pirate Boat, Castle with toys for children aged 3 to 6, etc. ªThere are also rollercoaster rides, an enchanted castle, a steam train and other attractions for all ages, such as the Army Museum, with military equipment, tanks and guns from World War II. There is also a small zoo in the park and a green house . http://www.thermashotelwalterworld. com.br
CITY OF BEES	It is located in Embu das Artes (SP). The park was created in the 1980s to introduce children and their families to the life of bees and the importance of this insect for all animal and plant life. It has an apiary, a giant beehive, a glass hive, a beekeeping museum, bee anatomy, bee panels, bouncy toys, slides and a honey house where you can buy bee products. The project entertains children and teaches them more about the construction of honeycombs in the hive. A fun, educational and

	entertaining attraction. You can see the bees producing in a glass hive with a real swarm. In addition to the entertainment, visitors can buy all the bee products: honey, honeycomb, propolis, pollen, royal jelly, honey bread and much more. In Embu das Artes there are signs indicating the location of the Parque Cidade das Abelhas. http://www.cidadedasabelhas.com.br
STATE OF IBITIPOCA	It is a state forest park located in the municipalities of Lima Duarte and Santa Rita de Ibitipoca, near the Mantiqueira mountain range in the southeast of Minas Gerais. It covers an area of 14.88 km^2 and has altitudes ranging from 1050 to 1784 meters. It has caves and water springs and several hiking trails. The soil contains a large amount of quartzite, which accumulates energy. It is administered by the Forestry Institute of Minas Gerais and was created in 1973. www.ief.mg.gov.br
	Thermas Park	Rio Verde (GO) http://www.silvestreresort.com/
	Beach Park	Fortaleza (CE) www.beachpark.com.br/
	Beto Carrero World Park	Balneario da Penha (SC) www.betocarrero.com.br
	Santa's Village	Gramado (RS) www.papainoel.com/
	Alegria Farm	Funilândia (MG) http://www.fazendaalegria.com .br
	Hot Park	Rio Quente (GO). http://hotpark.com
	Paradise Water Park	Porto Seguro (BA). http://www.arraialecoparque.com.br/
	Bird Park	Foz do Iguaçu (PR) http://www.parquedasaves.com. br
	Rainbow Falls	Ouro Fino-Ribeirao Pires (SP) http://www.rainbowfalls.com.br /
	Rio Water Planet	Vargem Grande-Rio de Janeiro (RJ) www.riowaterplanet.com.br/
	Venice Water	Maria Farinha-Paulista-(PE)

		http://venezawaterpark.com.br/
	Wet'n wild	Itupeva (SP) http://www.wetnwild.com.br/
	Parque Cidade do Vaqueiro	Manaus (AM) http://www.portalvaquejada.co m.br
	Mirabilandia Park	Salgadinho-Olinda (PE) http://www.mirabilandia.com.b r
	Beach Park	Porto das Dunas-Aquiraz- (CE) http://www.beachpark.com.br/
	Thermas dos Laranjais	Olimpia (SP) http://www.termas.com.br/
	Terra Magica Florybal Park	Canela (RS) www.parqueterramagicaflorvba l.com.br/
	Unipraias Camboriù Park	Balneario Camboriù (SC) www.unipraias.com.br/
For more information on each of these parks, see the websites indicated.		

4.5 Basic Tour Itinerary

Urban growth, the lack of options for contact with nature in the Metropolitan Region of São Paulo, and even on the coast, plus the violence in large urban centers, have caused people to move into shopping malls, cinemas and clubs.

Taking advantage of the natural conditions of Itatinga and aiming to offer new perspectives for ecological tourism in the region, this proposal for the Itatinga Ecotourism Theme Park - Itatinga EcoPark was developed.

It is hoped that visitors will enjoy nature and expand their knowledge of it.

The Basic Tour Itinerary complements the design of the Project and aims to describe the tour, through a preliminary interpretive analysis of this wonderful natural site, providing knowledge of plant and animal species, regional and local history, geography and pedology, biological processes and relationships, ecosystems, and the space of the Park itself, contributing to visitor awareness, an essential condition for environmental preservation and conservation.

As mentioned above, the main land access to the development site is from the Dr. Manuel Hyppólito Rego Highway (SP-55), entering the <u>Park's Main Access Sector </u>via Rua Manoel Gajo, which crosses the highway, and about 500 (five hundred) meters to the left after crossing the highway bridge over the Itapanhaù River, in the Santos-Sao Sebastiao direction.

Rua Manoel Gajo (formerly Vasco da Gama), belongs to the Secondary Road System of the Municipality of Bertioga.

Another option for accessing the park would be by sea, circling the island of Santo Amaro, entering the Bertioga Channel, and then the Itapanhaù River, or else, originating in the Port of Santos Estuary, sailing through the Bertioga Channel and also accessing the Itapanhaù River.

The options for access by ship would obviously require a complete restructuring of this mode of transport.

[a] In the Park's operational concept, this alternative was considered to be another option for the future and would be part of a **second stage of the Park**, as it would involve resizing the current mooring on the right side of the Itapanhaù River, in the area adjacent to the "Bondinho" boarding station.

In the **Park Sectorization** proposal, the entrance areas are located on land on Rua Manoel Gajo. These "entrance areas" would comprise the parking lot for charter buses, private cars and bicycles; an area for the Urban Public Transport System stop, maneuvers and spaces for road reversibility and the construction of the Entrance Gate, Reception and Control Equipment, Ticket Offices, etc.

The areas required for this sector have been indicated in <u>TABLE 1 </u>of this work.

After crossing the Portal, tourists and visitors will be taken to the *Port of Embarkation and Mooring* to cross the Itapanhaù River by boat.

Currently, Companhia Docas do Estado de Sao Paulo - Codesp has a boat to ferry employees and residents of the plant. When the Park is set up, it will be necessary to acquire a specific boat for the Theme Park, which will only be used to ferry tourists, visitors and users.

Aerial view of the crossing (IPECAB Collection)

The boat crossing of the Itapanhaù River, which takes 5 to 10 minutes, marks the start of the trail. Afterwards, you board a yellow-painted "cable car" and enjoy the exuberant scenery along the way, as you bounce leisurely along the rails for approximately 7 km. The scenery is made up of crystal-clear streams, sandbank vegetation on the plateau and the edges and escarpment of the Serra do Mar, covered by hillside forests.

The Itapanhaù River is associated with the local aborigines (Tapanhunos and Miramomis). Some historians associate the name of the river, originally ***Tapanhù***, with the existence on its right bank, between the river and the mountains, of villages of *Tapanhunos* Indians. On old maps and charts, the current Itapanhaù River was identified as "Tapanhù".

The Itapanhaù River was considered a tributary of the Bertioga River (now the Bertioga Canal). There are historical references to the existence of "farms" with sites for Jesuit catechesis. The *Jaguareguara or Jaguareguava and Pelaes* rivers, tributaries of the Itapanhaù River, were places dedicated to the conversion and social assimilation of the *Tapanhunos and Miramonis.*

The Itapanhaù River was important for the famous *Bertioga Quilombo*, led by "father Felipe", which existed near its tributary Jaguareguara (va) and was attacked and destroyed between 1835/1836.

The great Quilombo of Santos, formed around 1780, was located between the large hills of "Cabeça de Negro" and "Jaguareguara", which were entered from Bertioga, across the Itapanhaù River.[14]

The Itapanhaù River - and most coastal rivers - change their appearance after descending the mountains and reaching the lowland plateau. As they flow down the steep slopes of the

---

14 **HISTÓRIA DE SANTOS**, by Francisco Martins dos Santos 1st Edition. 1937.

mountains, their waters are crystal clear. This is because they only advance over bare rock, and they run with great speed, carrying gravel and stones. These gravels are also characteristic because, as they roll continuously, they take on rounded shapes, which is why they are called *pebbles*.

Once on the plateau, with no slope, the river slows down and only carries very fine sediments - clays, plant remains and silt that are deposited in its bed.

That's why, in the lowlands, the bottom of the rivers is not stony but muddy - often with a dark silt because of the plant remains brought in from the forest. This mud - as well as the humus-laden water from the forest floor - is very fertile and rich in nutrients.

These conditions interfere with the color of the river's water. As it moves away from the slopes of the mountain, it changes color, becoming dark and somewhat muddy.

Crossing the Itapanhaù River in the direction of Itatinga, these variations are clearly visible, especially near the crossing point. In addition, the areas adjacent to the river on this stretch are covered by mangrove vegetation.

The mangrove swamp, made up of twisted trees and rich in nutrients, is home to a variety of fish, prawns, lobsters, mullets, shellfish and a variety of other coastal animals.

The "Bondinho de Itatinga" station is located in the "Portinho de Itatinga" (former port of Fazenda Pelaes).

Port of Itatinga (IPECAB Collection)

The construction of the plant was authorized on August 22, 1905 (Decree 5.646), but soon faced enormous difficulties in transporting men and materials to continue the work. As a result, a 0.80 meter gauge railroad was built, taking advantage of the machines that operated on the old Santos pier (Kraus).

The railroad, stretching 7,250 meters from the port of Fazenda Pelaes (on the banks of the

Itapanhaù River) to the Itatinga Power Station, was laid alongside the path of the power transmission lines.

Started in 1905 and completed in mid-1906, the railroad began operating with the German steam engines of Fabril Kraus & Cia and Münchem & Linz (there is an example preserved in the Port of Santos Museum).

As the cost of steam traction increased, it was decided to electrify the line, which was considered a "streetcar line" because it ran within the then municipality of Santos (Itatinga was in the district of Bertioga - municipality of Santos. Bertioga was only emancipated in a plebiscite held in 1991); Itatinga was electrified in 1958.

An electric locomotive that had been working on the refurbishment of the line was transformed into streetcar No. 2. There is no information on the origin of this locomotive, but it may have been a Cia City "Progresso" locomotive, one of the first in the country.

Image of the cable car (CODESP Collection)

Operating from the beginning with two (2) streetcars, these were coupled to non-motorized trailers, which came from the old "donkey streetcars" of the Cia City de Santos, numbering three, which were refurbished and adapted in the workshops of the Cia Docas de Santos. The Itatinga Railway still operates in this way today.

Prior to the installation of the power station, village, railroad, etc., the area belonged to Fazenda Pelaes, located between the Itapanhaù River and the slopes of the Serra do Mar.

There are historical references to the former Fazenda Pelaes being a Jesuit catechesis center, with the participation of Fathers Leonardo Nunes, Nóbrega and Anchieta.

The old Fazenda is mentioned in historical documents as a refuge for pirates, forging coins, etc.

In 1925, Spanish farmer José Vergara was one of the pioneers in introducing banana cultivation, when he bought 1,800 plots of land between the Itapanhaù River and the slopes of the Serra do Mar, next to the Itatinga Power Station area. The banana plantations had trolleybuses pulled by animal traction, pushed by retailers and machines.

The bunches were cut and taken to the "chatôes", boats used for transportation, being towed by motorboat down the Itatinga-Jaguareguava-Itapanhaù-Canal de Bertioga rivers and finally to the Port of Santos, from where they were shipped to Argentina and Uruguay.

On some stretches and trails you can see the remains of old trolleybuses.

Image of a segment of the old trolleybuses (IPECAB Collection)

The streetcar departs from the station in the direction of Paredao da Serra and after about 400 meters from the port it passes by archaeological ruins and makes a sharp, almost 90-degree turn to the right; then, heading northeast, it heads towards the village of Itatinga.

The lanes in front of the tracks are covered with fragrant marsh lilies, bordering much of the approximately 7 km route.

On the left, you can see the "Wall" of crystalline rocks of the Serra do Mar, the lush hillside vegetation with trees 20-25 meters high and a wide variety of shrubs. Many of them produce flowers. In spring and summer, the Serra's landscape is transformed, offering the beautiful spectacle of blooming white, pink, purple and yellow flowers.

You can also find a wide variety of vines coiled around the stems and trunks of the tallest trees and countless orchids, with their showy flowers, such as lilies, catleias and stanopias; or smaller, delicate ones like the golden rains. Another beauty is the high-altitude escarpment and the crossing of numerous small, stony, crystal-clear streams cut by the streetcar tracks.

With patience and attention, you can observe the flight of birds, parrots and toucans.

On the right-hand side of the trail, there is vegetation known as "mata de restinga", a kind of transitional forest, where in addition to various species typical of the mountains, there is a tangle of trees, vines and often thorny shrubs, such as gravatâs (similar to pineapple trees, but much larger), as well as a wide variety of small coconut trees, tucum, brejaùva, a large number of lianas, bromeliads, guambês and countless species of palm trees, such as palmiteiros and jerivâs.

Both on the mountain side and in the transitional forest, along the route of the "cable car" you'll often see thickets of imbaùbas, with their showy, jagged leaves, always at the highest points.

This is the favorite plant of the Sloth, one of the representatives of the group of toothless mammals (the others are armadillos and anteaters). The "Sloth" loves to cling awkwardly to the high branches of larger trees and feed on the shoots and new leaves.

But it's not just the sloth that lives with the imbaùba. Among the ants, there is a species that has its only place of residence in the hollow of these trees. They build their nests there after drilling holes in the trunk to look for food outside.

At the base of the imbaùba leaves there is a kind of fluff, in the middle of which small sticks grow, similar to tiny grains of rice. The ants feed on this material.

Houses in Itatinga Village (IPECAB Collection)

After this fantastic journey, which lasts about twenty-five minutes, you arrive at the **village of Itatinga**. The small village, with a single wide street lined with gardens, is home to around 70 houses without walls or fences, all in the same color (green and yellow).

The village was built using English construction techniques and its design dates back to 1890. In addition to the great architectural value of the houses with their English style, we have the Chapel of Our Lady of the Conception built with some pieces that were dismantled and brought from England, including the altar, door and stained glass windows.

Our Lady of Conception Chapel - Itatinga (IPECAB Collection)

Nowadays, many homes are literally "closed", as the number of plant employees living there has decreased. Around a hundred people work at the plant, including Codesp employees, employees of the Cooperative of Workers of the Port Union of the State of São Paulo and freelance workers from the supplementary force.

Although not all the workers live in the village, Codesp still has a school building (inaugurated on June 20, 1918), a medical center and a small grocery store, as well as the

Itatinga Athletic Club and the 50-seat amphitheater (formerly a cinema), all set amid lawns, trees, foliage, flowers, seeds and birds.

At the end of the village you come to the Hydroelectric Power Station, a stone-walled building whose collection system, pipes, turbines and generators were designed by the engineer Guilherme Benjamim Weinschenk, whose work began in August 1906, after the railroad was started in 1905.

Inside the Power House, the floor is decorated with tiles, the window sills are made of granite and the railings are gilded with gold.

It was inaugurated on October 10, 1910, at the time with a nominal power of 20,000 KVA (20MW).

Main Facade of the Power House Building (IPECAB Collection)

Descent of the water tank to drive the plant's turbines (IPECAB Collection)

The water is collected from the Itatinga River at the top of the Serra do Mar, at a height of approximately 900 meters above sea level, at the bend of the Itatinga River, where there used to be a waterfall located close to the border between the municipalities of Bertioga and Mogi das Cruzes.

At this point, a dam was built, whose water was diverted through a stone gallery, dug into pure rock and about 800 meters long, to a point further down and close to the "edge" of the Serra do Mar, in a place known as the "Chamber of Equilibrium". At this point, known as: "the Chambers of Water", mark the beginning of the channeling in five pipes, which run down the mountainside to the five turbines connected to the plant's five generators.

The plant is currently operating with four of the five generators, producing 16 MW.

One of the turbines under maintenance (IPECAB Collection)

The power generated at the Itatinga Hydroelectric Power Station is taken to the Port of Santos via a 30 km long transmission network with five substations: Fazenda Caiubura-Bertioga, Caetê and Monte Cabrao (Santos mainland), Torre Grande (Vicente de Carvalho, in Guaruja) and Macuco (in Santos).

Vila de Itatinga, a small town nestled between the Itatinga River and the slopes of the mountain, has several waterfalls and small streams that help to embellish the landscape.

In this little paradise, without violence or crime, without robberies or assaults, without a trace of asphalt, nature dominates. A little higher up, on top of a small hill, amidst bamboo, palm and mango trees, where it can be seen from anywhere in the village, is the beautiful Chapel of Our Lady of the Conception, patron saint of the place, where everything is in perfect order and harmony: clean and painted walls and external staircases, an interior with an altar decorated with flowers, white lace tablecloths and a red carpet; colorful stained glass windows decorated with sacred themes, images of sacred art and small paintings reproducing scenes from the Way of the Cross.

From the village of Itatinga you can walk along several trails, enjoying the silence of the woods, listening only to the sound of the waterfalls in the distance and the singing of the many birds.

Most of the trails, with the exception of those that go up towards the mountains, are flat and easy to walk. You can reach the banks of the Itatinga River in about 30 minutes.

As you walk along the trails along the river, the dynamics of how these watercourses are shaped by the summer floods are striking. The stretches you walk along give you access to numerous natural pools with their clear, crystalline and cold waters, where you can watch the shoals of tiny colorful fish and take unforgettable refreshing swims.

This clean *water* is a testament to the preservation of the hillside forests in the Itatinga region, as it is this Atlantic Forest that filters out the dirt and other impurities that would cloud the river's *water*.

Occasionally there are moments of murky water, usually after heavy rains the previous few days.

The natural pools are full of pebbles, small "beaches" of coarse sand, gravel and small stones.

Crossing the river to the other bank, you can see the effects of the torrents up close, find animal footprints and, in the calmer stretches of the river, witness the magic of tadpole metamorphosis.

Although told as stories from the past, many things in Itatinga are worth remembering.

The stories of the Itatinga Atlético Clube festivities, where an Iambuzado pig was released for people to run after in an attempt to catch it; the races from the Usina to the portinho - along the cable car route, the games of sack race, egg race, pau-de-sebo and soccer matches.

He also remembers the large barbecues that were held with the donation of old José Vergara, who owned land in the area and gave an ox as a gift every year.

Night fell and the entertainment began with forró, the sound of good baiâo and old songs.

The lively June festivals and free movies, now a thing of the past, are mentioned.

In the sectorization of the Theme Park Project, the Village was included in the <u>Visitors Sector</u> and the trails in the <u>Trails and Viewpoints </u>Sector.

These are the areas for contemplation, resting and eating, the sports area, the special use area and other park equipment and products.

CHAPTER 5 - PROPOSALS

5.1 Proposals that involve and depend on the Public Authorities Municipal for PARK SUPPORT

Making the Park compatible with Integrated Municipal Planning, Bertioga's Sustainable Development Master Plan and the Tourism Development Sector Plan (Tourism Master Plan).

Recommendations and Necessary Actions:

• Review of municipal zoning, establishing new standards for the use and occupation of urban land in areas close to the park's main access.

• Definition of new restrictive urban planning standards in the areas adjacent to the main access.

• Strictly control the slumification of the areas around the access to the Park.

• Improve the urban transport service by increasing the number of lines and timetables to Portal do Parque.

• Improve the public cleaning system, especially around the park's main entrance.

Implement a Municipal Urban and Tourist Signage System.

Recommendations and actions needed: *Install signposts indicating the park along the access roads to the municipality, especially on the Dr. Manoel Hyppólito Rego highway (SP-55).

Public Works to Support the Park: *Construction of the Passenger Terminal for Urban Public Transport.

*Construction of a cycle path along Rua Manoel Gajo to the "Portal do Parque".

*Installation and construction of a roof for a police vehicle to support the park's security.

*Installation of the sewage system on the access road to the park and rainwater drainage works.

*Improve street lighting on the main access road to the park.

5.2 Proposed responsibility of the developer for each sector of the Park

5.2.1 For the Main Access Sector

► Construction of the Access Gate to the Park (Name of the Park) in Rua Manoel Gajo.

► Construction of a building next to the Gate for the <u>Ticket Office, Reception, Security and Information Center,</u> for ticket sales - communication - orientation - control and information.

► Construction of male and female toilets in the Portal area.

► Implementation of a private demarcated area for <u>closed and controlled parking</u> of private cars, bicycles, motorcycles and charter buses, with a guard booth - control - communication - gate and fire prevention and fighting equipment.

► Implementation of a <u>Passenger Terminal for Urban Public Transport,</u> with a boarding and disembarking point, with cover, lighting, identification of timetables, itineraries, benches, etc.

► Definition and dimensioning of maneuvering areas and road reversibility.

► Expansion of the covered area for passengers (tourists) at the <u>Port of Embarkation</u> and construction of male and female toilets, including benches for sitting.

► Urban and support furniture: public telephone booths, garbage cans, mailboxes, drinking fountains, toilets, benches, equipment to prevent and rescue people from boating accidents (launching buoys, cables, motorized boats to support the crossing, etc.).

► Adequate lighting in the Portal and Portinho areas.

► Acquisition of a boat suitable for crossing the Itapanhaù River, sized and equipped to cater for tourists, including access for people with physical disabilities and other special needs;

► Adaptation of the embarkation and disembarkation berths, with the construction of a roof and appropriate ramps.

5.2.2 For the Route and Rail Sector

► Expansion of the facilities at the "<u>Bondinho Station</u>", including toilets, benches for sitting and waste collectors for "selective collection".

► Demarcation of the <u>archaeological site</u> along the route, with cleaning of the surroundings and illumination of the ruins.

► Installation of <u>information, tourist and environmental signs</u> along the "cable car" route.

► Define one or two intermediate stopping areas for contemplation and photos.

► Extend the coverage of the arrival point in the village, including benches, toilets and drinking fountains.

► Build a closed electric tramway for any aggressive weather: storms and gales.

► Define the maximum passenger capacity for crossing the river and traveling to the village.

5.2.3 For the Main Visitors' Sector (Vila)

<u>Support and administration equipment</u>:

► **HOME OF THE MONITOR:**

*House designed to house the park's monitors, with a place to shower, change clothes, rest and store material.

* Keep a stock of raincoats, umbrellas and rubber boots.

* Keep an antiophidic serum, a First Aid kit and equipment for removing the injured.

* Maintaining the radio, TV and mobile telephone communication system.

* Keep a storeroom stocked with flashlights, batteries, tools, etc.

5.2.4 Administration House for the General Administration and <u>THEMED HOUSES</u>[15] of the Park:

► <u>HOUSE OF THE SEA</u>: It will be organized to show marine life, oceans, aquatic organisms, the seabed, naval heritage, preservation of the sea, research at sea, etc. It will contain photographic panels, images and a small *Museum of the Sea.*

► <u>CASA DO ÌNDIO</u>: Will be organized to show the history of the Indians in Brazil, in the coastal region, in the Baixada region and in Bertioga. It will contain panels, photos,

---

15 THEMED HOUSES: Making use of houses that are currently unoccupied and others that may be vacated, to add a collection with a specific theme, without de-characterizing the architectural aspects.

handicrafts and all the demonstrations of indigenous life, customs, celebrations and traditions. It will be decorated with motifs from indigenous culture.

► HOUSE OF FISHING: It will be organized to tell the story of fishing, fishing tackle, replica boats, maps, material used for fishing in different eras, old and new equipment, nets, species, etc.

► FOREST HOUSE: Will be organized to show in detail the characteristics of the Atlantic Forest, Restinga Forest, Coastal Plain Vegetation, Mangroves and other National Biomes.

► ENERGY HOUSE: This will be organized and set up to tell the story of energy in the world, in Brazil, in the region and at the Itatinga Hydroelectric Power Station. It will contain a historical collection of the Itatinga plant, documents, photos, maps, books, small objects, etc. The *Energy Museum* is attached.

► PIRATES' HOUSE: This will be organized to tell the story, legends and habits of the pirates who used to attack Brazilian shores. It will be set up with objects, clothes, weapons, equipment, etc. The presence of pirates in the region will be highlighted, as well as the history of their struggles and expulsion.

► HOUSE OF WATER: It will be organized by showing the importance of water in the world, in Brazil and in the region. It will show the relationship between water and the history of the country, its importance for life, ways of protecting rivers and springs, etc. Including a small aquarium, mainly with the region's ichthyofauna (fish from rivers and streams in the park area, as a thematic element for environmental education). A GIANT PLUVIOMETER should be installed in front of the WATER HOUSE for educational purposes.

► CASA DO JESUíTA: Will be organized to tell the story of the arrival of the Jesuit priests in Brazil, in the region and in Bertioga. The importance of the Jesuits in the country's education, indigenous catechesis, the protection of mixed-race children, works, legends, tales, miracles, etc....

The work of Fathers José de Anchieta, Leonardo Nunes, Manuel da Nóbrega and others will be highlighted.

► HOUSE OF BLACK CULTURE: Will be organized to show the importance of black people in Brazil. The African continent, black culture and black history, slavery, slave ships, black labor, the Quilombos of the region, black leaders, etc.

5.2.5 Seedling Nursery, Bromeliad and Orchid Nursery

*Seedling Nursery: A place to organize a seed bank of native trees and produce seedlings.

*BROMELIA AND ORCHIDARIA: A place for research, production and development of bromeliads and orchids, especially native ones.

5.2.6 Other proposals for the visitor sector:

*INN: Small inn, if possible, using the original houses of Itatinga village.

*LANCHONETTE: Structure for quick meals and snacks.

*PLACE FOR CHILDREN'S RECREATION: Containing instructional and recreational toys plus places for games.

*CONVENIENCE SHOP: Selling souvenirs and handicrafts, caps, T-shirts, gifts, postcards, posters, paintings (canvases), books, sanitary pads, diapers, repellent, etc.

*SPORTS AREA: Use of the facilities of the *Itatinga Atlético Clube.*

*TROUBLE AND CLOTHES: For use by tourists and visitors.

*RESTING PLACE WITH SMALL DRINKING WATER FOUNTAIN.

*AREA WITH STAGE FOR THEATRICAL AND MUSICAL PERFORMANCES.

*CENTRAL AREA OF THE VILLAGE WITH PARK MAP AND TOURIST GUIDE.

*INSTALLATION OF WASTE COLLECTORS FOR THE "SELECTIVE COLLECTION" SYSTEM.

*KIOSKS: For rest, conversation and contemplation.

5.2.7 For the Trails and Viewpoints sector

The trails and viewpoints sector aims to make interactive use of natural resources and attractions.

The sector is responsible for "exploring" and enhancing natural beauty, clean air, water, landscapes, plants and animals.

The trails and viewpoints sector, from the point of view of the physical environment, includes all the trails and paths around the village of Itatinga.

PROPOSALS:

☐ <u>Indication and mapping of all the trails,</u> including signposting, preventive measures against accidents, structuring them for ecotourism use (including educational signposting) and setting up interpretive programs.

☐ Create a viewpoint at a <u>height</u> that allows you to see the Itapanhaù River basin, the meeting of the Itatinga River with the Itapanhaù, the sky and the sea.

☐ Create a special trail, called TRILHA DA FAZENDA, in agreement with the local owner, Mrs. Eliana Vergara, in order to take advantage of the structure and area of the farm, as well as to introduce the theme of the banana plantations that once existed.

5.2.8 Recommended bans, of a preventive and preservationist nature

1. Prohibit access to and transportation of weapons within the Park.

2. Prohibit the chasing, catching, slaughtering, trapping, hunting or capturing of birds and other wild animals in and around the Park.

3. Prohibit the use of any type of explosive.

4. Prohibit the removal, collection, gathering or scavenging of specimens of flora or fauna, including roots, seeds, leaves, stems, resins and fruit.

5. Prohibit religious or non-religious celebrations that involve the throwing of any material or object into springs or watercourses (streams, creeks, rapids or waterfalls), into the forest, with the emission or generation of any type of waste or particulate matter (bonfires and fires), smoke, etc.

6. Prohibit the dumping of garbage or other materials that pollute the soil, *water* or air, or compromise the landscape, health or scenic integrity of the Park.

7. Prohibit the installation or fixing in the areas of the Park of signs, siding, notices or signs that are not related to the work of the Park or to the preventive, safety or interpretive programs.

8. Prohibit any form of camping in the Park areas.

5.3 Proposal for Stage 2ª Itatinga Eco Park II

1. Study the use for tourist purposes of the inclined plane, parallel to the network of water pipes of the current *winch system*, with steel cable traction, and the existing tracks.

2.   Use and structure the trail along the "*Stone Channels*" from the Water Chamber to the Itatinga River dam at the top of the mountain.

3.   Construction of an "Observatory Lookout" in the area near the Water Chamber, at the height of the mountain.

4.   The Natural Bath House would be built and implemented in the village of Itatinga, in an area close to the soccer field, taking advantage of the clean, crystal-clear water of the Itatinga River and channeling the current discharge from the hydroelectric turbines. The bathhouse would have a hydrotherapy character .[16]

5.   Build an Alternative Covered Area (multiple uses), for rainy days, in a suitable location, without causing deforestation or landscape impact, considering the high rainfall rates in the area.

---

16 **HYDROTHERAPY:** Referring to Hydrotherapy, which consists of treatment using water in external applications, baths, sprinkles, showers, vaporization, etc. Hydrotherapy dates back to the ancient Romans who considered it an efficient means of healing, hygiene and recovery from fatigue (stress).

CHAPTER 6 - FEASIBILITY AND SUSTAINABILITY STUDIES

6.1 Park numbers

In order to draw up feasibility projections for the Park, it was based on market potential assessment studies carried out by the Research, Analysis and Business Consulting Company Toledo & Associados at the end of the last decade, and on studies by SEBRAE .[17]

Data and information from the Baixada Santista Development Agency - AGEM, Empresa Paulista de Planejamento Metropolitano S/A - EMPLASA and the Brazilian Institute of Geography and Statistics - IBGE.

The studies considered the areas of influence of the Park, i.e. the Metropolitan Region of São Paulo, the Metropolitan Region of Baixada Santista and the Metropolitan Region of Campinas (SP).

6.1.1 Attendance[18]

Há various methodologies for calculating "Attendance". The preliminary reference number is taken as the basis for market estimates.

It usually works with an "Area of Influence".

As the distance from the business increases, the number of people willing to go there decreases. Therefore, the ***concept of distance*** is used as a delimiter of the public. Based on North American assumptions, a theme park has a radius of influence of a maximum of 160 km in a straight line; therefore, the following regions were considered for the purpose of this study: Baixada Santista Metropolitan Region, Sao Paulo Metropolitan Region and Campinas Metropolitan Region.

- <u>Resident Population in the Metropolitan Region of Baixada Santista</u> (made up of the nine municipalities of: Bertioga, Cubatao, Guaruja, Itanhaém, Monguagua, Peruibe, Praia Grande, Santos and Sao Vicente).

Total: <u>1.8 million inhabitants (IBGE 2017/EMPLASA)</u>[19]

- <u>Resident Population in the Metropolitan Region of Sao Paulo</u> (made up of 39

---

17 **SEBRAE:** Brazilian Microenterprise Support Service
18 **ATTENDANCE:** English term, used in the planning and management of North American theme parks, which means "the capacity of the market" to have people interested in attending the development, or the capacity of the market to offer a certain presence of visitors. Potential Audience.
19**EMPLASA:** Empresa Paulista de Planejamento Metropolitano S/A - website: www.emplasa.sp.gov.br

municipalities, including the capital of the State of Sao Paulo).

The municipalities are divided into sub-regions and Sao Paulo is part of all of them:

<u>Sub-Regions and Municipalities:</u>

North: Caieiras, Cajamar, Francisco Morato, Franco da Rocha and Mairipora.

East: Arujà, Biritiba-Mirim, Ferraz de Vasconcelos, Guararema, Guarulhos, Itaquaquecetuba, Mogi das Cruzes, Poà, Salesópolis, Santa Isabel and Suzano.

Southeast: Diadema, Mauà, Ribeirao Pires, Rio Grande da Serra, Santo André, Sao Bernardo do Campo and Sao Caetano do Sul.

Southwest: Cotia, Embu, Embu-Guaçu, Itapecerica da Serra, Juquitiba, Sao Lourenço da Serra, Taboao da Serra and Vargem Grande Paulista.

West: Barueri, Carapicuiba, Itapevi, Jandira, Osasco, Pirapora do Bom Jesus and Santana de Parnaiba.

Total: <u>21.3 million inhabitants (Population Estimate)</u>

<u>IBGE/2016/EMPLASA)</u>

- <u>Resident Population in the Interior of the State of São Paulo</u>

Area of Influence of the Development (Metropolitan Region of Campinas), (Formed by 20 municipalities).

<u>Municipalities</u>: Americana, Artur Nogueira, Campinas, Cosmópolis, Engenheiro Coelho, Holambra, Hortolândia, Indaiatuba, Itatiba, Jaguariûna, Monte Mor, Morungaba, Nova Odessa, Paulinia, Pedreira, Santa Bàrbara d'Oeste, Santo Antônio de Posse, Sumaré, Valinhos and Vinhedo.

Total: <u>3.1 million inhabitants (IBGE Population Estimate 2016/EMPLASA)</u>

Thus, for the year 2016/2017 (IBGE/EMPLASA estimate, 2016/2017) the **TOTAL POPULATION OF THE AREA OF INFLUENCE CONSIDERED IS:**

26.2 million inhabitants

or

26,200,000 people

Based on this data, two methodologies were used to arrive at the <u>potential target audience </u>for the development:

1)   Cut-off method by region and income;

2)   Penetration Rate Cutting Method.

1) Cut-off Method by Region and Income

This is a population cut-off by average family income, based[20] on IBGE data, called the Target Income Rate.

This study used a rate of 19.47%[21] for a population with the ideal income profile to visit the park, taking into account transportation costs, food, etc. Therefore:

- Population in the area of influence: 26,200,000

Target income population = 19.47%

(19.47%) of 26,200,000 = 5,101,140

People with income in the Area of Influence = 5,101,140 people

2)   Penetration Rate Cutting Method

This methodology is widely used in the United States and seeks to determine the maximum distance a visitor is willing to travel to reach the park.

Considering the Embraparque studies, as indicated in the footnote, three regional zones of influence were selected:

1ª) ZONE = Formed by the cities of the Baixada Santista Metropolitan Region.

2a) ZONE = Formed by the cities of the Sao Paulo Metropolitan Region.

3a) ZONE = Formed by the cities of the Metropolitan Region of Campinas -SP

Thus, adjusting the population in the area of influence to the average penetration percentages gives the following situation:

- Total population of the Target Area = 26,200,000

20TARGET INCOME RATE: A term used in the United States to refer to the population with a "target income".
21The 19.47% figure was used as a "Target" in the Embraparques Theme Park Project - Itanhaèm - SP, under the guidance and criteria of the Toledo & Associados - Pesquisa, Anàlise e Consultoria company. Therefore, for academic purposes, it was decided to apply the same number in this work.

Average population <u>penetration rate</u>[22] = 18.22%[23]

(18.22%) of 26,200,000 = 4,773,640

Target audience calculated according to IBGE 2016/17 data = 4,773,640

Applying the Simple Average of the Methods used =

Method (1) Method (2)

5.101.140 + 4.773.640 = 9.874.780 ÷ 2 = 4.937.390

Target = 4,937,390 people

Following the same methodology and after defining the number of people who make up the Park's target audience, the <u>intention to attend</u> should be verified through market research.

<u>Market coverage</u>: this is a multiplier that looks at the percentage of people who will actually be able to buy a product or service, in relation to the <u>Target Audience</u>.

According to the studies *already* mentioned, even taking market coverage into account, it was adopted:

Potential Park Audience (Park Audience) = 1,530,000/year

It is important to note that this potential number, known as "Park Attendance", does not yet take into account the <u>turnover factor</u> (number of annual visits by each visitor) and the <u>seasonality factor,</u> which causes attendance to vary (reduction factors) according to the different times of the year (school vacations, seasons... spring, summer, autumn, winter): spring, summer, fall, winter), variations in the country's economy, unemployment rates, the specific characteristics of each theme park, the cost of tickets, etc.

Both factors are decisive for the size of ATTENDANCE.

6.1.2 Turnover Factor

Turnover is defined as the number of times a visitor goes to the park over a period of one (1) year.

In order to obtain the "<u>Adjusted Factor</u>", it is advisable to carry out market research that is

---

22 **PENETRATION RATE:** This is the percentage of people willing to travel the distance from each Zone of Influence to the development site. Again, the rates from the research work for EMBRAPARQUE were used. Itanhaém Park - SP.
23 **18.22%:** This is the weighted average of the Penetration Rates applied in the respective zones of influence (1st, 2nd and 3rd ZONES, i.e. RMBS, RMSP and RMCAMP).

always up-to-date and aimed at the Target Audience.

In Embraparque's studies, an average of 3 (three) times a year for passport holders and 1 (one) time a year for the "Day Park" was adopted as the turnover assumption.

It's worth noting that these figures are well above the averages for American parks.

According to the results of the aforementioned FADE-OMT-EMBRATUR survey carried out in Brazil, the average annual number of visitors to Theme Parks is 33% of their capacity, reaching 53% in the high season (percentage in relation to the total number of enterprises in Brazil).

The best months are the high season months: January, February (summer) and July (winter school vacations).

According to the same survey, the break-even point for theme parks is around 18%, which means that they can already achieve non-negative net results.

This level of activity is considered to be quite reasonable, even though it is less than 40%, because the operational capacity of the resorts envisages a growing future demand, which may benefit from economic stabilization and other conditioning factors, such as the national and international trend of increasing demand for leisure.

It should be noted that the data from the surveys used to obtain the average annual number of visitors to theme parks in Brazil predates the economic crisis that the country has been facing since 2015, with unsatisfactory economic results and high unemployment rates. Clearly, the country's political and economic situation has a direct impact on visitor attendance at theme parks.

6.1.3 Seasonality Factor

Seasonality is a well-known phenomenon for coastal businesses. It's an effect that usually causes numerous operational and financial problems. Generally, demand for services and products in the coastal region increases significantly in the summer season, when Brazilians traditionally spend a few days in coastal towns to "enjoy" the beach, sun and sea.

Estimates indicate that on the coast of the state of São Paulo, "beach" resort towns suffer an average increase of 3.5 to 4.0 times in their fixed population. This is known as the "floating population".

This demand caused by the floating population creates problems in the distribution of electricity, gas, drinking *water*, supplies, excess and congestion on highways and urban roads, access to public services (health, security, etc.).

One of the objectives of the seasonality analysis is to build an ***Attendance Curve*** for the park during the year, checking how they will be distributed month by month.

There are various methods for assessing the ***seasonality of*** a region or a city. It is possible to construct several graphs and then cross-reference them to obtain consistent average values.

Some examples:

*   Emergency medical service (Emergency Room) attendance data.

*   Sales variation data in local commerce.

*   Electricity consumption data.

*   Bank transaction data.

*   Variation data on urban public transport users.

*   Vehicle data (cars, vans and buses) at the toll stations on the coastal highways and in other ways.

*   Fuel consumption data at gas stations in the supply network.

The months with the highest projection of the "Attendance Curve" will be the months with the highest heat and, consequently, vacations, which are = December, January and February, as well as a small increase in July (due to the winter school vacations). It is also possible to evaluate the concentration of the main holidays and the interference they can produce in that month.

In water theme parks, the months of May and June are usually closed to the public and the period is used for staff vacations and preventive maintenance of the equipment.

6.2 Carrying Capacity

ITATINGA ECOPARK = "Attendance" + Support Capacity

Although knowledge of these methodologies is important; and, in the case of the Itatinga Ecotourism Theme Park, the analysis makes it possible to demonstrate market viability, it is the **park's carrying capacity** that will define the average number of visitors and daily visitors.

There is a lot of research and studies underway in various countries around the world, trying to establish qualitative and quantitative criteria for sustainability in tourist areas and developments.

In National Parks, it is the "Management Plans" that will present the explicit estimates for the carrying capacity of each area exposed to visitation.

In turn, the "Management Plans" of National Parks and other Conservation Units work on the adequacy of the Visitation Rate (meaning the number of annual visitors per unit of area and used to assess the frequency of visitors).

Basically, visitor entry to the Parks depends on the size of the Park. There is a limited flow rate and, if managed, the rate can be permanent. The sustainable income of a Park depends on the entrance fee.

However, carrying capacity is an ecological concept; therefore, it is not just a quantification of the maximum number of visitors a park can receive, but **how much it has been preserved, revitalized and educated**.

The carrying capacity is the maximum number of visitors a park can receive over a given period of time without showing obvious signs of degradation, such as: soil erosion, loss of soil along excessively trodden trails, dumping of waste in preservation areas, suppression of vegetation by users, capture of animals, pollution of watercourses, etc.

Proper management of the Park implies keeping resources in balance with flows. The size and "ecological health" of the Park will lie in the protection of the soil, water, air, vegetation, fauna, and all the factors that feed the Park.

Nowadays, there are experiments to estimate the number of visitors that can travel along certain trails (these values decrease in areas with rugged topography). For example, caves and caverns don't support as many visitors as the trails leading to them, and so on. Mapping and detailing all the trails is essential.

The aim of sizing the carrying capacity is to obtain a sustainable return on flow rates and on the Park's structure and resources.

Generally, in private ecotourism ventures, entrepreneurs try to maximize profits, disregarding strict control over the number of visitors.

Itatinga EcoPark's proposal is fundamentally ecological and preservationist, always considering environmental values above pecuniary values and interests. The Theme Park concept considers it possible to obtain financial results and profitability, as long as the entrepreneur manages the business with a commitment to planning, marketing and administrative efficiency.

[a]In the case of the *Itatinga Ecotourism Theme Park*, in order to define the carrying capacity of the 1st stage of the project, we started with a historical analysis of the plant and the community that has lived there for the last hundred years. The current structure of Itatinga and the constraints of the project itself have shown us what standards of visitation are admissible with full capacity to be supported.

The basic references for calculating the quantification took into account the plant's daily routine and the number of buildings in the area:

BRIEF DESCRIPTION	QUANTIFICATION
Number of houses built and installed in the village since its foundation. Type of construction: Single-family house.	70 (seventy)
Number of houses built in areas contiguous to the village, the way up the mountain and construction at the top of the mountain.	15 (fifteen)
Total number of residential buildings.	85 (eighty-five)
Brazilian average number of inhabitants per family.	5 (five)
Total number of people to live in the plant.	85 x 5 = 425
Estimated average number of people attending per day, worked or visited the village (teachers, doctors, priests, technicians and engineers, suppliers, freelancers and visitors, family or institutional).	80 (eighty)
Total number of permanent residents or occasional visitors (daily average).	505 (five hundred and five)
Current number of people working, living or visiting the plant.	150 (one hundred and fifty)
SO 505 - 150 = 355 PEOPLE/DAY	

An analysis of the above data shows that with <u>355 people/day</u>, the park will be able to absorb the existing structure.

For the purposes of defining operational standards, the average number of employees in Brazil's theme parks is 0.02 - 0.03 per visitor (FADE/EMBRATUR survey).

355 x 0,03 = 10,65

Considering the operational characteristics of the Itatinga Theme Park, a number up to 3 (three) times higher could be accepted. It would be:

10.65 x3 = 31.95 → 32 employees

Reducing the number of employees by 32 from the total number that can be supported

355 - 32 = 323

323 would be the base number to guide the **Park's Management Plan**, or by definition: <u>Admissible Carrying Capacity</u>.

If you consider that the current cable car, with trailer, can comfortably carry 60 (sixty) people:

- Streetcar → 6 benches with 4 seats = 24 seats and

- Trailer → 9 benches with 4 seats = 36 seats

60 seats

This flow will be distributed as follows

355 people day ÷ 60 = 5.38 trips/day

Considering a journey time of approximately 40 minutes (round trip) we will have 5.38 x 40' ≅ 215' =

3h and 58' or 4 hours x 2 maximum flow

Assuming the Park's <u>operational planning</u>, it will be possible to organize convenient departure and return times.

Continuing the operational analytical exercise and adopting the average annual value of tickets in Theme Parks in Brazil (FADE/EMBRATUR) as:

US$ 22,22

Permitted carrying capacity of 323

323 x US$ 22.22 = US$ 7,177.06/day

Considering that the average annual number of visitors to theme parks in Brazil is 33% of their capacity (data already mentioned),

Transforming the estimate   We used the exchange rate: 1US\$ = 3R\$

The average revenue per day would be **R\$ 6.459,00**

The number achieved only reinforces the economic and financial viability of the project.

Of course, the financial profitability of the venture can reach higher levels. This *"Plus"* will depend on: the partnerships to be formalized by the Entrepreneur, the Business and Management Plan, the Management Capacity, and the Marketing Plan.

6.3 Environmental Impact Assessment

It is not the purpose of this work to detail and evaluate environmental impacts.

Impacts are defined as the significant alteration of existing environmental conditions.

The Itatinga Ecotourism Theme Park Proposal does not propose any deforestation or physical intervention works.

There will be no significant changes in the 1st^a Stage of the Park. Not even the flow of people to the Itatinga Power Station will be altered.

As explained above in the sizing of the bearing capacity, we tried to maintain the original living characteristics of the site.

During the implementation phase, a broad program of participation by the Bertioguense community is considered important, clarifying, disseminating and making people aware of the importance of *Itatinga as a historical, cultural, technological and environmental heritage site* and its value as an increaser of sustained tourism in the municipality.

Of course, in the operational phase, there are a number of potential impacts, many of which have been considered in advance and resolved once the proposal has been adapted.

In the 2nd Stage of the Project, scheduled for the 2nd three-year period of the Park's operation, the manager will have obtained sufficient experience and knowledge of the operational procedures and implemented the Environmental Monitoring Program provided for in the **Management Plan** for the Park area.

^{a}On this occasion, the Planning, Design and Implementation Department will have the technical means to assess, minimize and control any possible impacts of the operational phase and prepare the studies for the implementation of the second stage of the project, called

Itatinga EcoPark II.

6.4   Implementation stages

The proposed development was planned to be implemented in two successive stages:

ia STAGE - ITATINGA ECOPARK I (1st Triennium)

- Public support for the Park (already mentioned in Chapter 5).

- Access structure (Portal - Reception - Information, etc.).

- Refurbishment of the port and mooring.

- Furniture.

- Tourist boat for crossings.

- Refurbishment and adaptation of the station.

- Park signage.

- New reserve cable car.

- Themed houses.

- Seedling nurseries, bromeliad and orchid nursery.

- Snack bar.

- Convenience store.

- Recreation and sports areas.

- Fountain.

- Amphitheater.

- Kiosks.

- Collectors and garbage.

2ª STAGE - ITATINGA ECOPARK II (2nd Triennium)

- Inclined plane.

- Trail of the stone canals.

- Elevated observation deck.

- Natural baths.

☐   Alternative multiple-use coverage.

☐   Hostel.

6.5   Considerations on the Management Plan

This work does not aim to build a Management Plan for the Park, but recommends the following:

6.5.1 Management

The entrepreneur must organize at least **four managers**:

1.   Planning and Project Management

Duties: - Managing the plans, projects and implementation of the Park.

- Manage programs and actions.

- Developing new ideas.

2.   Operations and Maintenance Management

Duties: - Managing the operation of the park.

-     Managing the general upkeep of the park.

- Manage security.

3.   Administration and Finance Management

Duties: - Raising revenue.

-     Control expenses.

-     Managing costs and investments.

-     Managing staff, etc.

4.   Marketing Management

Duties: - Developing the Marketing Plan.

- Manage and control marketing actions.

6.5.2 Security Concept

It is very broad, as it encompasses a variety of situations and states of both an objective and subjective nature. Generally speaking, security is defined as the absence of unforeseen events, threats or risks.

Basically, security must know the vulnerabilities and risks so that preventive measures can be taken to ensure that everything goes as planned and desired.

Security must be integrated into the **Management Plan** and there must be no ostentatious demonstration of security systems and teams. Many developments have internal CCTV, preferably "camouflaged" in the park environment. Security must not detract from the visitors' sense of leisure and relaxation. As far as the operationalization of security is concerned, two things are recommended: rapid identification of the event (still in its infancy) and prompt response.

Theme Parks have large and grandiose areas, many attractions and large concentrations of people; therefore, they need permanent monitoring.

6.5.3 How the parks work

For illustrative purposes only, the following shows the shape and opening hours of some Theme Parks in Brazil.

	Park Name	Days/entry	Timetables and Values
OPERATION	Fazendinha Park in São Paulo - Estaçâo Natureza	2ª and 3ª	10 a.m. to 5 p.m.
		Sat/Sun and public holidays	10 a.m. to 5 p.m.
		Entry	Adults and children R$ 45
		Parking	R$ 15
	Park Beto Carrero Consult the website *www.betocarrero.com.br*	Every day (See calendar)	8:30 a.m. to 7 p.m.
		Single passport (which varies depending on the time of year)	On average, from R$77 to R$100/day.
		Children over 10 pay the adult price.	
	Aparecida Park (Magic Park)	Thursdays	10 a.m. to 6 p.m.
		Fridays/Sat/Sun and holidays	9 a.m. to 9 p.m.
		Adults = R$ 14,90	

		Children between 86 cm and 1.20 cm tall = R$ 8.90	
		Children up to 85cm in height are free.	
		>65 years = R$ 9,90	
		People with disabilities, free admission.	
	New Parque da Mônica (Always check the website: *www.parquedamonica.com.br*	During school hours, only on weekends.	
		Friday	10:30 a.m. to 4:30 p.m.
		Sat/Sun and public holidays	10am to 7pm
		Half price	R$ 72,50
		Individual	R$ 145
		Packages:	
		2 people	R$ 190
		3 people	R$ 282
		4 people	R$ 372
		5 people	R$ 460
OPERATION	Hopi Hari Park (Opening hours outside the vacation period). On vacation, check the website: *www.hopihari. com.br*	Fridays	10 a.m. to 6 p.m.
		Sat/Sun	10 a.m. to 6 p.m.
		Parking	Value not available.
		Adult passport	R$ 99
		Children's passport	R$ 49,50
		Some attractions have different prices, for example, the Katakumb costs R$ 20/per person, the Tirolesa costs R$ 25/per person.	
		People over 65 free of charge.	

BIBLIOGRAPHIC, ICONOGRAPHIC REFERENCES[24] AND PHOTOGRAPHIC REFERENCES

1) <u>Secretariat for the Environment of the State of Sao Paulo</u>. *Cadernos de Educaçâo Ambiental - Concepts for Environmental Education.*

Environmental Education Coordination. 39 ed. Sao Paulo, 1999.

2) <u>State Secretariat for the Environment</u>. *Education, Environment and Citizenship - Reflections and Experiences*. Environmental Education Coordinator. Various authors. Sao Paulo (SP), 1998. 122p.

3) <u>Ministry of Education and Sports</u>. *The Implementation of Environmental Education in Brazil*. Publication of the Coordination of Environmental Education. Brasilia- DF, 1998.

4) <u>IBAMA - Brazilian Institute for the Environment</u>. *Education for a sustainable future - A transdisciplinary vision for shared action*. IBAMA edition: Brasilia, DF, 1999.

5) <u>Bee Theme Park </u>- Sao Paulo: Sao Paulo. Available at: <u><http://www.cidadedasabelhas .com.br/</u>

6) <u>Ibitipoca State Park</u>. Minas Gerais (MG). Available at: <u><http://www.ief.mg.gov.br</u>

7) <u>Thematic</u>. Available at: <http<u>:www.parques.bol.com.br/temâticos/index.jhtm</u>

8) <u>BRASILIANO,</u> Celso Ribeiro. *The Concept of Security for Theme Parks*. (Graduated in Business Administration, former EB-AMAN officer, specialist in Security Systems and Director of COPS - Consulting Division.

9) <u>SAO PAULO -SP - BERTIOGA.</u> Sustainable Development Master Plan - PDDS. Municipal Law No. 316, of November 13, 1998.

10) The <u>Electrification of the Itatinga Hydroelectric Railway -</u> Electrification on Brazilian Railways. 2003. Available at: <http://www.efbrasil.eng.br/electro/hi.htlml

11) <u>The Itatinga Power Station - History</u>. 2003.

Available at: <u>>http://wwwgeocities.com/ferrovias rasil/ita- historia.htm</u>.

---

24 ICONOGRAPHY: relating to iconography. (1) Description of knowledge of images, portraits, paintings or particularly ancient monuments. (2) Art of representing by means of images. (3) Knowledge and description of images (engravings, photographs, etc.). (4) Visual document that completes a work of reference and/or of a biographical, historical, geographical nature, etc.

12) NUNES, Nilo. Itatinga Ecotourism Theme Park. Part I and Part II. Issues 347 and 348 - Sao Paulo: 1999. Costa Norte Newspaper. June 12/18 and June 19/25 - Environment Column.

13) Master Plan Project Coord. Presentation - Introduction - Methodology and Foundations. Bertioga City Hall. 1994 e 1995.

14) BRAZILIAN ASSOCIATION OF GASTRONOMY AND HOSPITALITY AND TOURISM - Thematic Parks in Brazil. Available at: < http://www.abresi.com.br/htmls/parques tematicos.htm

15) ITALY -VENEZA - *Venice Charter*. II International Congress of Architecture and Historic Monument Techniques. May 25-31, 1964.

16) SAO PAULO. *Office GP-1076/00 of July 13, 2000 - Case No. 40.036/00.* CONDEPHAAT - Council for the Defense of the Historical, Archaeological, Artistic and Tourist Heritage of São Paulo.

17) GDMDS Order No. 137/00, of July 10, 2000.

18) RESOLUTION No. 237, of December 19, 1997. National Environmental Council - CONAMA. Disciplines and Regulates Environmental Licensing.

19) FEDERAL LAW No. 9.735, April 27, 1999, provides for Environmental Education and other measures.

20) RESOLUTION No. 001, of January 23, 1986. CONAMA. Establishes general guidelines for the use and implementation of Environmental Impact Assessment.

21) FEDERAL LAW No. 6.938 of August 31, 1981, provides for the National Environmental Policy, its purposes and mechanisms for formulation and application, and makes other provisions.

22) FEDERAL LAW No. 7.661, of May 16, 1988. Establishes the National Coastal Management Plan and makes other provisions.

23) FEDERAL LAW No. 9.985, of July 18, 2000. Establishes the National System of Conservation Units and makes other provisions.

24) FEDERAL DECREE No. 3179/99, which regulates FEDERAL LAW No. 9.605, of February 12, 1998 - the Environmental Crimes Law.

25) FEDERAL DECREE No. 84.017, of September 21, 1979, Approves and Regulates

Brazilian National Parks.

26) <u>FEDERAL DECREE No. 99.274,</u> of June 6, 1990, provides for the creation of Ecological Stations and Environmental Protection Areas.

27) <u>RESOLUTION No. 275,</u> of April 25, 2001. CONAMA - Establishes the color code for different types of waste, to be adopted in selective waste collection programs.

28) <u>SAO PAULO - STATE ENVIRONMENTAL SECRETARIAT.</u> RESOLUTION - SMA No. 77, of November 24, 1997. Resolves to define developments intended for leisure and recreation, such as the so-called "Thematic Parks", in terms of the requirements of SMA RESOLUTION No. 42/94.

29) <u>EMBRATUR</u> - Economic and Financial Study of Hotels and Theme Parks in Brazil. Brazilian Tourism Studies. Brazilian Tourism Institute. September/1999. Chapter IV, pages 183 to 205.

30) <u>AMUSEMENT PARKS AND ATTRACTIONS ASSOCIATION</u> - World Association of Theme Parks and Attractions.

31) <u>ANDRADE, M.A.B.; LAMBERT, A. A.</u> Vegetation. In: AZEVEDO, A. (Ed.) A Baixada Santista. Sao Paulo: EDUSP, p.151200, 1965.

32) <u>NEW MÔNICA'S PARK</u> - Website:<http://www.monica.com.br/parque/shop old/welcome.htm

33) <u>MAGIC PARK APARECIDA.</u> Website:

http://www.amadeusturismo.com.br/parques/parques T/Nac/Magic P.htm

34) <u>HOPI HARI PARK.</u> Website: < http://www.hopihari.com.br (Sao Paulo)

35) <u>VIVA PARQUE.</u> Website: http://www.vivaparque.com.br/

36) <u>SKI MOUNTAIN PARK.</u> Website: http://www.skipark.com.br

37) <u>WATERWORLDPARK.</u> Website: http ://www.thermashotelwaterworld.com.br

38) <u>ADIBRA</u>: Associação das Empresas de Partes de Diversoes do Brasil.

39) <u>Website:</u> < http://www.venezawaterpark.com.br (Pernambuco)

40) <u>Website:</u> < http://www.betocarrreroworld.com.br (Balneàrio de Penha - SC).

41) <u>Website:</u> < http://www.parquedogugu.com.br (Sao Paulo - SP)

42)   Website: < http://www.beachpark.com.br (Ceara)

43)   Website: < http://www.terraencantada.com.br (Rio de Janeiro - RJ)

44)   Website: < http://www.thermaspark.com.br (Rio Verde - GO)

45)   Website: < http://www.playcenter.com.br (Sao Paulo and Pernambuco)

46)   Website: < http://www.skimountainparque.com.br (Sao Roque-SP)

47)   BERTIOGA-SP. Photos. Photo collection of the Bertioga Environmental Sciences and Research Institute - IPECAB.

48)   RAP: Embraparque, Volume I and II. Itanhaém- SP, February/1998.

49)   ENVIRONMENTAL MACROZONING OF THE NORTH COAST - Bertioga District. Report and Diagnosis. Volume I. Ins. Engineering and Consulting and Management, Coastal Planning Division of the Environment Secretariat of the State of São Paulo, December/1990.

50)   SANTOS AND REGION: Guide to Professional Events. Santos and Region. Convention & Visitors Bureau.

51)   NATIONAL TOURISM PLAN: Guidelines, Goals and Programs (2003-2007). Ministry of Tourism. Brasilia, April 29, 2003.

52)   ECOTOURISM IN ITATINGA. Route and interpretation - Preliminary version. Santos City Hall and CODESP. Various authors. 1992.

53)   BRANCO, Samuel Murgel. *The Serra do Mar and the Baixada*. Sao Paulo: Ed. Moderna, 1992.

54)   GERIBELLO, Denise Fernandes. The patrimonialization of industrial structures: the case of the Itatinga Mill. Doctoral thesis. FAUUSP. Sao Paulo, 2016.

Printed by Books on Demand GmbH, Norderstedt / Germany